Studies in Computational Intelligence

Volume 1205

The series "Studies in Computational Intelligence" (SCI) publishes new developments and advances in the various areas of computational intelligence—quickly and with a high quality. The intent is to cover the theory, applications, and design methods of computational intelligence, as embedded in the fields of engineering, computer science, physics and life sciences, as well as the methodologies behind them. The series contains monographs, lecture notes and edited volumes in computational intelligence spanning the areas of neural networks, connectionist systems, genetic algorithms, evolutionary computation, artificial intelligence, cellular automata, self-organizing systems, soft computing, fuzzy systems, and hybrid intelligent systems. Of particular value to both the contributors and the readership are the short publication timeframe and the world-wide distribution, which enable both wide and rapid dissemination of research output.

Indexed by SCOPUS, DBLP, WTI AG (Switzerland), zbMATH, SCImago.

All books published in the series are submitted for consideration in Web of Science.

Mikołaj Morzy

Spoken Language Processing

Conversational AI for Spontaneous Human Dialogues

Springer

Mikołaj Morzy
Faculty of Computing
and Telecommunications
Poznań University of Technology
Poznań, Poland

ISSN 1860-949X ISSN 1860-9503 (electronic)
Studies in Computational Intelligence
ISBN 978-3-031-88565-5 ISBN 978-3-031-88566-2 (eBook)
https://doi.org/10.1007/978-3-031-88566-2

This Springer imprint is published by the registered company Springer Nature Switzerland AG
The registered company address is: Gewerbestrasse 11, 6330 Cham, Switzerland

So, while speech and writing can both be very complex, the complexities tend to be of different kinds. The complexity of speech is choreographic—an intricacy of movement. That of writing is crystalline—a denseness of matter. In linguistic terms, spoken language is characterized by complex sentence structures with low lexical density; written language by simple sentence structures with high lexical density. We could express this even more briefly, though at the cost of distorting it somewhat, by saying that speech has complex sentences with simple words, while writing has complex words in simple sentences

—M. A. K. Halliday

Preface

There is a growing need to establish a new scientific domain dedicated to the processing of spoken language, a field that would focus on developing techniques that enable intelligent machines to understand and respond to spontaneous human speech. A significant number of innovations in Artificial Intelligence (AI) and Machine Learning (ML) is required in order to bridge the gap between human communication and machine understanding. The objective is not merely to decode the semantics of spoken language but also to understand the nuanced elements such as context, sentiment, and intent. Spontaneous human speech is difficult. It is unplanned and unrehearsed. Unlike written language, which is often carefully crafted and revised, spontaneous speech is usually impromptu and does not follow a predetermined structure. Spoken language is full of fillers, hesitation markers, and non-verbal cues such as tone, pitch, volume, and pace, which provide additional context. A lot of meaning is conveyed complementarily to speech, for instance, using pauses and interruptions, which can carry semantic and pragmatic meaning. Transcriptions of spontaneous speech may resemble natural language, but it is so different from the prescriptive written language, that current tools and models trained on huge corpora of written text suffer from significant performance degradation when applied to spontaneous speech.

The book primarily focuses on the expansive field of Spoken Language Processing (SLP). It begins by introducing the unique characteristics of spontaneous human speech, describing the nuances and complexities that make it distinct from written language. It then provides an overview of the current state of Conversational AI (CAI), shedding light on recent advancements and ongoing research in this domain. The book presents key technologies instrumental for spoken language processing, including Automatic Speech Recognition (ASR) and transcript engineering, which form the backbone of understanding and interpreting spoken language in a machine-readable format. To aid in practical application and research, the book includes appendices that provide a list of training datasets and benchmarks.

The primary audience for this book are researchers in the field of Natural Language Processing (NLP) looking to expand their horizons into new and challenging domains of application. The experience in working with prescriptive written text can be a valuable asset when entering the territory of spontaneous human conversations. Bridging the gap between NLP and SLP is an important task that can unlock many research possibilities and this book aims to be the guide for such explorations. The second group targeted by this book are students of artificial intelligence and machine learning, who want to explore beyond text and images and venture into the domain of multi-modal learning on audio and text. Finally, the book is designed with a specific focus on those interested in exploring the innovative ways in which AI can assist in tasks involving spontaneous human-to-human communication. The book explores how AI can be used in the development of virtual assistants, chatbots, and other conversational AI systems that mimic the dynamics of human conversation. By examining these applications, the book aims to inspire readers to think creatively about how AI can transform the way we communicate, making the process more efficient, inclusive, and engaging.

Obviously, the field of spoken language processing is vast, combining the fields of artificial intelligence, machine learning, deep learning, computational linguistics, speech recognition, speech synthesis, human-computer interactions, phonetics and phonology, acoustics, not to mention psychology and cognitive sciences. Necessarily, the author had to make a choice of the main topic of focus and had to significantly reduce the scope of the subject matter covered in the book. Among the subjects not covered in the book are the details of natural language processing (word embeddings, sentence embeddings, document embeddings) and descriptions of neural network architectures popular in NLP (recurrent neural networks, long short-term memory networks, gated recurrent unit networks). The book does not focus on the popular Transformer architecture and does not go deeper into the details of large pre-trained generative models forming the backbone of the ongoing NLP revolution. The depth of presentation of the ASR technology and ASR model training had to be limited as these subjects would require separate monographs to cover them adequately. Linguistic, phonetic, and acoustic issues are presented briefly and in the context of how they are affected by the characteristics of spontaneous human speech.

The main aim of this book is to serve as the starting point in the journey into the fascinating field of artificial intelligence for spontaneous human speech. The book focuses on the features of spontaneous speech that make it such a formidable challenge for contemporary AI models. However, a word of caution is due. The field of AI is advancing at such a rapid paste that it seems every freshly conceived book on AI is already outdated before the ink is even dry on the manuscript. The author has poured his soul into describing the latest breakthroughs and developments in spoken language processing, only to find that his work is superseded by new revelations before the book even reaches the press. The dynamism that makes AI so exciting is the same force that renders the author's efforts to document it a Sisyphean task. It is a bitter pill to swallow, this constant race against obsolescence. But such is the life of those who choose to plunge into the ever-evolving world of artificial intelligence.

Poznań, Poland
September 2024

Mikołaj Morzy

Acknowledgements

In 2016, I found myself at a crossroads when I lost my secondary employment at a local college. It was a challenging period, and I was not sure what the future held. However, fate had an unexpected twist in store for me when one of my former students reached out to me with a proposal to join a startup. The idea intrigued me: collaborating with an American company specializing in call center applications and infrastructure. Although I was initially hesitant, I decided to give it a go, and little did I know that it would lead to a wild adventure spanning over six years. This former student of mine (and now my boss) was Jan Mizgajski, a man of hundred talents (of which artificial intelligence stood out), who kept pushing me far outside of my comfort zone. In this one person, I have got a very demanding supervisor, a keen debater on all topics AI, and a warm-hearted friend. Without Jan, not a single word of this book would have been written.

Balancing my work for the startup with my administrative and teaching duties at the university was no easy task. I found myself constantly juggling responsibilities and pushing the boundaries of my superior's patience. But despite the challenges, it was an exhilarating experience that allowed me to grow both personally and professionally. I managed to convince one of my best master students, Adrian Szymczak, to quit his job at Amazon and return to Poznań to join the startup. It was a moment of immense satisfaction to see him thrive, learn, and radiate magic positivity on everything and everyone he came into contact with. For all 5 years of our work, Adrian was tormented by the rest of the team about the mythical length of his master thesis. I hope Adrian will not treat my confession as a treason: his master thesis was longer than this book.

One day when I walked into our office, I have met a young gentelman politely smiling at me. Piotr Żelasko left Cracow and moved to Poznań with his family. A talented trombone player and the member of the Krakow Street Band, he knew everything about designing and training ASR systems. As it turned out, his only flaw was an annoyingly noisy keybord producing clickety click sounds, which he stubbornly refused to exchange for something more socially acceptable. It is safe to say that everything I know about ASRs, I have learnt from Piotr, and he was always generous with sharing his knowledge and vast expertise.

Not without pride, I note that I was quite instrumental in convincing Jan to hire two young and extremely talented Ph.D. students from Wrocław. Piotr Szymański and Łukasz Augustyniak joined our team and traveled to Poznań on a regular basis. I am probably not able to fully express how much I owe them, but I will try: countless hours of discussions on natural language processing and artificial intelligence, hundreds of suggestions for valuable research papers, dozens of tools and libraries (to this day I sometimes manage to impress my current students with my knowledge of the AI toolbox, and I am just parroting what I have learnt from Piotr and Łukasz), several papers that we have written and submitted together, hours of complaining about the ossified, outdated, and corrupt structure of Polish universities, and a lot of radical social commentary.

Over the years, the team grew and we were constantly looking for new talent. For some strange reason, almost all of our hires were my former students. Krzysztof Surdyk brought very solid expertise, a lot of stoicism, and very positive vibes. Wojciech Mikołajczyk was like *Joysinia Pestis*, a bacteria of optimism infecting everyone within seconds with good spirits and joy in this rare, non-iritating way. It was like some magic ability, he would be explaining the structure of the code he has written and all would be thinking "*Gosh, am I lucky it is raining today, I love the rain so much!*".

Jan, Adrian, both Piotrs, Łukasz, Krzysztof and Wojciech formed our machine learning team. Opposite us stood the "infra team", a fearful pack of hardcore coders. Bartosz Borowik was the best programmer I have ever met. My suspicion is that one could ask Bartosz to start coding in archaic sumerian and, after 1 week, Bartosz would be back with a compiler and a *HelloWorld* example. Robert Głowski was another freakishly talented programmer, who would divide his time equally between epic coding and heated cultural and philosophical polemics. Łukasz Wójciak, the front-end virtuoso later turned project manager, somehow managed to cycle an average of 1000 km per day without his JavaScript ever missing a beat. Marcin Baran, our back-end developer, alternated between coding and hardcore exercises followed by ice baths. His irony levels were definitely off the charts, no one could ever tell if Marcin was joking or being serious. This was particularly worrisome for me as at this level of coding sophistication I hardly understood what he was doing or talking about. Adam Artajew was our tester, but looking at his wits and sharp humor I think he could easily win bread by doing standup. The defensive line of the infra team consisted of devops. Every sane person in IT is afraid of devops. Our devops were no exception. Daniel Smoczyk, Adam Wróbel, Daniel Fryc, and Adam Lis could solve all of your problems given that you were worthy of their time and your problem was not a trivial one. Which produced something of a chicken-and-egg problem, because almost every problem a mere mortal could have was trivial for them, ergo, by having this problem one was demonstrably too stupid to waste their time. But as an academic, they treated me with special care (even when my stupidity and incompetence must have been causing them physical pain).

The third part of the startup was the data annotation team, consisting of linguists: Marzena Żyła-Hoppe, Joanna Banaszczak, Marta Muszyńska, Karol Kaczmarek, and Piotr Anusiewicz. They taught me to truly appreciate the baffling beauty of language, the complexity of language, and the elusive nature of language understanding. Without them, my understanding of language would be purely technical. They instilled in me a much deeper reverence for the language (and goth culture).

For 6 long years, we were working together, at first in the beautiful old house in the heart of Poznań, later, confined by the global pandemic to our appartments and seeing each other through the lenses of out laptop's cameras. Whatever the circumstances, it was pure joy and an endless learning opportunity. This book is the summary of everything I have learnt from these young, talented people. Later, in the book, I will precisely relate chapters and sections to our joint papers and I will attribute each contribution to individual people. But here I want to thank you all for taking me with you in this fascinating journey. I dedicate this book to all of you, my dear friends.

Last, but not least, I want to thank my wonderful family. Dagmara, Agata, and Marta create our microcosmos each day; it is where I feel loved, where I grow, and where I find my place in the universe. And there is one more person who is the real reason this book has been written. How many times can one answer, "*Yes, Grandma, I am just sitting down to write the monograph,*" "*Sure, Grandma, the book is in the writing,*" or "*Of course, next year at the latest I will bring you the book*"? Apparently, I have reached the limit of my Grandma's patience, so here you go, my dear Grandma, this book is especially for you.

Finally, this book owes its existence to my Father. A professor at the Poznan University of Technology, he was the one who instilled in me a fascination for computer science and a deep love for the rigors of academic work. From him, I learned what it means to be a scientist. He was the first reader of every chapter, a harsh but brilliant critic whose insights and relentless encouragement were instrumental in shaping this manuscript from its earliest stages. Tragically, he passed away just before he could hold this finished book in his hands. It is with deep sorrow that I dedicate this work to his memory, and I can only hope that he would have liked the book.

Contents

Acronyms

AI	Artificial Intelligence
AIML	Artificial Intelligence Markup Language
ASR	Automatic Speech Recognition
BOW	Bag of Words
CAI	Conversational Artificial Intelligence
CNN	Convolutional Neural Network
DNN	Deep Neural Networks
FST	Finite State Transducer
GMM	Gaussian Mixture Model
GRU	Gated Recurrent Unit
HMM	Hidden Markov Model
LLM	Large Language Model
LSTM	Long Short-Term Memory Network
MFCC	Mel Frequency Cepstral Coefficients
ML	Machine Learning
NER	Named Entity Recognition
NLP	Natural Language Processing
RLHF	Reinforcement Learning with Human Feedback
RNN	Recurrent Neural Network
SLP	Spoken Language Processing
SLU	Spoken Language Understanding
TFIDF	Term Frequency—Inverse Document Frequency
TTS	Text-to-Speech

List of Figures

List of Tables

Part I
Human Dialogues

Spontaneous human dialogues are characterized by their non-linear, dynamic, and unpredictable nature. This structure is vastly different from the structured and predictable pattern of written language or scripted dialogues. In spontaneous speech, interlocutors often interrupt one another, switch topics abruptly, use fillers and pauses, or even leave sentences incomplete. The conversation flow is also influenced by various factors such as the participants' emotional states, background knowledge, cultural context, and non-verbal cues like body language and tone of voice. Besides, spontaneous dialogues often include slang, colloquialisms, and idioms, which add another layer of complexity.

The unpredictability of spontaneous human dialogues makes them difficult to process, particularly for AI systems. AI models typically rely on pattern recognition, structured data, and pre-defined rules to understand and generate responses. However, the lack of structure, the presence of ambiguities, and the heavy reliance on context in spontaneous dialogues pose significant challenges. For instance, the same phrase can have different meanings in different contexts, and an interruption or sudden topic switch can completely change the direction of the conversation. Understanding non-verbal cues, cultural nuances, and emotional undertones, which play a crucial role in human dialogues, is still a largely unsolved problem in AI.

Chapter 1
Introduction

1.1 Why Spoken Language is Underrepresented in AI?

Language acts as the most fundamental and instinctual medium for human interaction and expression. It serves as a universal tool, free from the constraints of technology and not exclusive to any specific group. Unlike technological advancements that may be inaccessible to certain populations due to geographical, economic, or social barriers, language is a shared heritage of all humanity. It transcends beyond borders, cultures, and societies, enabling individuals to convey thoughts, emotions, and ideas, building understanding and connection. Regardless of the thousands of languages that exist, the capacity to communicate through language is a ubiquitous aspect of the human experience.

The origins of human language are steeped in history and mystery, and it is considered one of anthropology's most elusive questions. There exists a number of theories about the origins of language, including the gestural theory, the music theory, and the protolanguage theory. The gestural theory suggests that language originated from gestures, while the music theory suggests that it originated from music. The protolanguage theory suggests that language evolved from a simpler system of communication. According to [2], the probable age of human spoken language is around 150 000 years. The evidence supporting this theory includes the discovery of fossils of early humans with the vocal cords and brain structures necessary for speech, as well as the archaeological evidence of early humans using tools and symbols. In contrast, the earliest known form of written language, the cuneiform script developed by the Sumerians, dates back to around 3200 BCE [10]. The earliest examples of cuneiform writing are found on clay tablets from the Sumerian city of Uruk. These tablets are inscribed with pictograms, which are simple pictures that represent words or concepts.

M. Morzy, *Spoken Language Processing*, Studies in Computational Intelligence 1205,
https://doi.org/10.1007/978-3-031-88566-2_1

The advent of spoken language significantly predates written communication, highlighting the primary role of verbal interaction in the evolution of human societies. Unfortunately, this seniority of spoken language is not reflected in the contemporary AI space, where written rather than spoken language is used predominantly to train models. It is not surprising to see the research on natural language processing constrained to the domain of written text. Written text is far more accessible and abundant. The internet is a vast repository of written text in the form of books, articles, blogs, and social media posts that can be easily collected and used for training models. The written text provides a more structured and consistent form of language, while the spoken language is highly dynamic and unstructured. Additionally, spontaneous speech often includes interruptions, fillers, slang, and colloquialisms that are less common in written language, making it harder for AI to understand and process. Different speakers may have differing accents, pronunciations, and other idiosyncrasies introducing further obstacles.

The increasing demand for AI models trained on spoken language stems from the realization that communication is not merely a matter of words but also involves a rich tapestry of sounds. This includes obvious elements such as tone, pitch, pace, and emphasis, which can significantly alter the meaning of spoken utterances, but information is also encoded in pauses, hesitations, and omissions. These subtle audio cues often convey critical information about speaker's emotions, intentions, or even cultural background. If AI models can learn to understand and interpret these nuances, they would be able to provide a more accurate and human-like understanding of spoken language. Verbal human communication is important in numerous fields, including customer service, mental health, and online education. By enabling more empathetic, intuitive, and effective communication, AI models capable of capturing both textual and auditory information can make a step towards truly intelligent and human-friendly systems.

1.2 Conversational AI

The enduring promise of natural interaction with technology using human language is a vision that has captivated scientists, engineers, and users alike for decades. The idea of being able to communicate with machines in the same way we converse with fellow humans is not just appealing, but it could fundamentally transform our relationship with technology. For many years the leading AI field aiming to address this goal was NLP. NLP develops algorithms that can understand, generate, and respond to human language in a meaningful way. Since NLP is traditionally biased towards written text, the last few years witnessed the rise in popularity of the term Conversational AI. This term encompasses the set of technologies for processing both written and spoken language in the context of dialogues, either between humans or between a human and a machine. Chatbots, voicebots, home assistants, call center assistants, all these AI systems are becoming more advanced, understanding the literal meaning of words and sentences, and grasping the subtleties, nuances, and context of human language. This could lead to a future where interacting with devices is as easy and intuitive as talking to a friend: where digital assistants can understand and act on complex requests, where customer service bots can handle any query, and where home assistants can truly understand human needs, making them not just tools, but partners in our daily lives. Chapter 2 is devoted to the history of conversational systems, from rule-based chatbots, through statistical chatbots, to voicebots and chatbots powered by Large Language Models (LLM).

From the point of view of language processing technology, spoken language can be divided into two intertwined areas: automatic speech recognition and spontaneous dialogues. The latter presents an additional layer of complexity for AI systems compared to single-speaker spoken language. Spontaneous conversations often involve overlapping speech as multiple speakers may talk at the same time, which is a challenging scenario for speech processing systems to handle. The informal and unstructured nature of spontaneous conversations can often include a lot of filler words, repetitions, false starts, and utterances that serve only the purpose of turn control. The lack of clear punctuation markers and sentence structures in spontaneous speech complicates the tasks of sentence segmentation and punctuation restoration. We will discuss all the challenges of spontaneous dialogue processing in Chap. 3.

1.3 Speech Recognition

Every system for speech processing has to be able to represent the contents of the speech in some form suitable for further processing. Automatic Speech Recognition (ASR) refers to the technology that converts spoken language into written text. This task is complex due to the variability in speech across different individuals, accents, languages, and even the same person at different times. Background noise, overlapping speech, and speaker emotions further complicate the conversion. On top of that, the artifacts of spontaneous conversations (back-channeling, turn-taking, dialogue flow control) introduce much noise into the input audio signal.

The development of ASR models has gone through several significant stages over many decades. The history of ASR began in the 1950s with the Bell Labs' Audrey system, which could recognize spoken numbers. By the late 1960s, IBM's Shoebox machine was capable of understanding 16 English words. The 1970s saw the application of statistical methods, introducing Hidden Markov Models (HMM) which could handle larger vocabularies and continuous speech [4]. The 1980s and 1990s brought more advancements in the use of HMMs, leading to the development of large-vocabulary, speaker-independent continuous speech systems. Since the 2000s, with the advent of deep learning, ASR models have undergone substantial improvements. Deep Neural Networks (DNN) replaced Gaussian Mixture Models (GMM) [9] for acoustic modeling, leading to a significant drop in error rates. More recent advancements include recurrent neural networks and transformer-based models that have further improved the efficiency of ASR.

However, despite all recent advances in automatic speech recognition, the technology to accurately transcribe and annotate spoken language, a necessary step for training AI models, is not yet perfect. Significant strides were made in English and a few other major languages like Mandarin, Spanish, and French, largely due to the abundance of data available in these languages to train complex models. The same level of progress is not seen for the majority of the world's languages. This is primarily due to the lack of sufficient annotated data required to train and fine-tune models. Many languages, particularly those spoken by smaller populations or in developing regions, are under-resourced in terms of data availability. As a result, ASR technology for these languages is still in its infancy. Factors such as dialectal variations, lack of standardized orthography, and low digital presence, only exacerbate the problem. In recent years several techniques have been proposed that try to support under-resourced languages, mostly using transfer learning, data augmentation, and synthetic data generation, but the problems are far from being solved. A more thorough description of the field of automatic speech recognition and the presentation of state-of-the-art models is presented in Chap. 4.

The ultimate output of an ASR system is the transcript of processed utterances. However, producing a high-quality transcript faces numerous challenges that extend beyond the basic transcription of spoken words. Transcript engineering is a key aspect of refining raw transcripts produced by ASR systems, as it enables more accurate and contextually relevant transcriptions of spoken language. Typical steps in the transcript engineering pipeline include transcript annotation, transcript alignment, speaker diarization, and keyword spotting. Transcript annotation involves tagging elements of the transcript, such as speaker turns, emotions, or specific events. Transcript alignment refers to the process of synchronizing the transcript text with the corresponding speech in the audio, which is required for applications such as closed captioning and language learning. Transcript annotation is the practice of adding metadata or extra information to transcriptions of spoken text, often to highlight specific elements such as emotions, pauses, or linguistic features. This process is crucial for training and refining AI models, helping them to better understand and generate human speech patterns. Diarization is the task of identifying and segmenting different speakers in a conversation, which is particularly challenging in multi-speaker environments and is essential for understanding the context of a conversation. Segmentation is the task of identifying semantically meaningful spans of transcript that can approximate sentences in written language. Another task is true casing, where the model must correctly identify and apply the case (upper or lower) of the transcribed words, a task complicated by the fact that spoken language does not inherently contain information about capitalization. Yet another challenge is sentence segmentation — determining where one sentence ends and another begins, which is straightforward in written language due to punctuation, but considerably more difficult in spoken language, where pauses are not always indicative of sentence boundaries. Punctuation restoration is also a complex problem for ASR. In spoken language, commas, periods, question marks, and other punctuation aspects are often implied through pauses or changes in tone, but these cues can be subtle and vary widely between speakers. Each of these tasks contributes to enhancing the accuracy and utility of

ASR systems, making transcript engineering a key component in the development of speech recognition technology. A detailed description of tasks involved in transcript engineering is presented in Chap. 5.

1.4 Spoken Language Processing

The state-of-the-art NLP models are truly impressive and they are getting better (and much bigger) with every month. The same cannot be said of the parallel field of spoken language processing. Despite very promising advancements observed in recent years (such as the development of Whisper [8], a model from OpenAI specifically designed for speech recognition and translation tasks and trained on a massive dataset of audio and text covering over 50 languages), we still need significant research into spoken language processing. For instance, intent discovery from transcripts of spontaneous conversations is the task of identifying semantically meaningful sections where a speaker expresses their intention, request, or question. This task is challenging because there are numerous taxonomies of intents, each with its own nuances and categorizations. Furthermore, accurately identifying these intents is crucial for businesses, as it helps them understand customer needs and improve service delivery. The complexity of natural conversation, with its informal language, interruptions, and context shifts, adds another layer of difficulty to the task, making intent discovery a critical yet error-prone process.

Named Entity Recognition (NER) is the task of identifying and classifying tokens into predefined categories such as names of persons, organizations, locations, expressions of times, quantities, percentages, etc. While this task has seen significant progress in written text, it can be particularly challenging when applied to transcripts of spoken language. One major challenge is the informal and unstructured nature of spoken language which often includes variations, colloquialisms, mispronunciations, and non-standard grammar. This can make it difficult for NER models to accurately identify and classify entities. Another challenge is the lack of context in spoken language, as people often rely on shared knowledge or visual cues when speaking, which are not captured in transcripts. Correctly recognizing spans containing named entities is further complicated by homophony, the phenomenon of different words sounding the same (producing "*eye phone*" instead of "*iphone*" in the transcript). In general, word errors introduced by the ASR into the transcript usually break the grammar of text and are very "unnatural" from the point of view of NER models. Thus, despite advancements in NER, its application to spoken language transcripts remains an unsolved task. We discuss intent discovery and named entity recognition in spontaneous speech, as well as the characteristics of NER errors, in Chap. 6.

Spoken Language Processing (SLP) has a wide array of potential applications that can significantly impact society. In the realm of personal computing, virtual assistants like Amazon's Alexa, Google Assistant, and Apple's Siri use spoken language processing to understand and respond to user commands, making technology more accessible and intuitive. In the business sector, SLP can be used in customer service centers for call analytics, sentiment analysis, and training purposes, improving the overall customer experience. It also has applications in healthcare, where it can be used to transcribe and analyze patient-doctor conversations, aiding in accurate diagnosis and treatment. In the field of education, SLP can be employed in language learning apps, enabling users to learn new languages more effectively. It can also be used to facilitate accessibility for individuals with hearing impairments, by providing real-time transcriptions of spoken language. In legal and law enforcement contexts, SLP can be utilized for transcribing court proceedings or police bodycam footage, ensuring accuracy and accountability. In entertainment, it is used for automatic subtitling and dubbing of movies and TV shows. In the transportation industry, SLP can support hands-free control of vehicles, contributing to safer driving. As technology continues to advance, the potential applications of SLP are likely to expand and evolve, shaping the way we interact with the world.

While the book provides an extensive overview of various topics around SLP, there are some key concepts related to NLP and ASR that are not covered in detail. This includes the use of pre-trained language models, such as BERT [3], GPT-4 [6], or Llama 3 [5], which have revolutionized the field of NLP in recent years. The book also omits the more basic discussion of word and sentence embeddings, both static and contextual. Recurrent Neural Networks (RNNs) and Long Short-Term Memory (LSTM) networks [7], designed to handle sequential data like text or speech and used extensively before the advent of the Transformer architecture [11], are not elaborated upon. We do not cover in detail the Transformer architecture as well. The book also does not discuss the technical aspects of ASR training or the use of wav2vec [1], a self-supervised model that can be used to improve the performance of ASR systems. The book does not discuss the field of phonetics, which studies the physical sounds of human speech and is critical for understanding the intricacies of spoken language. These omissions are due to the book's focus and scope, more specialized resources for a deeper understanding of these topics are provided as literature references.

References

1. A. Baevski, Y. Zhou, A. Mohamed, M. Auli, Wav2vec 2.0: A framework for self-supervised learning of speech representations. Adv. Neural Inf. Process. Syst. **33**, 12449–12460 (2020)
2. G. Deutscher, *The Unfolding of Language: An Evolutionary Tour of Mankind's Greatest Invention* (Macmillan, 2005)
3. J. Devlin, M.-W. Chang, K. Lee, K. Toutanova, Bert: Pre-training of deep bidirectional transformers for language understanding. arXiv preprint arXiv:1810.04805 (2018)
4. B.H. Juang, L.R. Rabiner, Hidden markov models for speech recognition. Technometrics **33**(3), 251–272 (1991)
5. L. Van Der Maaten et al., The llama 3 herd of models. arXiv preprint arXiv:2407.21783 (2024). Accessed 9 Sep 2024
6. OpenAI, Gpt-4 technical report. Available at: https://arxiv.org/abs/2303.08774 (2023). Accessed 9 Sep 2024
7. J. Oruh, S. Viriri, A. Adegun, Long short-term memory recurrent neural network for automatic speech recognition. IEEE Access **10**, 30069–30079 (2022)

8. A. Radford, J.W. Kim, T. Xu, G. Brockman, C. McLeavey, I. Sutskever, Robust speech recognition via large-scale weak supervision, in *International Conference on Machine Learning* (PMLR, 2023), pp. 28492–28518
9. D.A. Reynolds et al., Gaussian mixture models. Encyclopedia Biomet. **741**, 659–663 (2009)
10. D. Schmandt-Besserat, The evolution of writing. Int. Encycl. Social Behav. Sci. 1–15 (2014)
11. A. Vaswani, N. Shazeer, N. Parmar, J. Uszkoreit, L. Jones, A.N. Gomez, Ł. Kaiser, I. Polosukhin, Attention is all you need. Adv. Neural Inf. Process. Syst. **30** (2017)

Chapter 2
Conversational AI

2.1 What is Conversational AI?

The simplest definition of Conversational AI states that it is a type of artificial intelligence that enables humans to interact with computer applications the way they would interact with other humans. Another definition stresses the fact that CAI is the synthetic brainpower that enables machines to understand, process, and respond to interactions expressed in human language. CAI goes beyond the exchange of messages, it refers to the use of artificial intelligence to facilitate human-computer interactions that are natural, intuitive, and intelligent. This involves understanding and interpreting human language, generating appropriate responses, and even anticipating and responding to follow-up questions or statements. While CAI draws on NLP, it goes beyond this field in terms of the focus on user experience, stateful dialogue management, multi-modality (text and audio), and context management. CAI incorporates machine learning and deep learning algorithms to facilitate dynamic and meaningful interactions. It learns from past conversations, understands the nuances of human language, and improves its responses over time, providing a more interactive, personalized, and engaging user experience. In essence, while NLP forms the basis for language understanding, CAI elevates this foundation to facilitate real-time, contextual, and meaningful dialogues.

CAI offers several benefits to both businesses and users. By automating routine tasks, it helps save significant amounts of time, allowing businesses to focus on core competencies and users to receive immediate responses. CAI can assist customers in making purchasing decisions by providing personalized recommendations based on customers' preferences and browsing history. Businesses can also benefit from the potential to sell outside of traditional business hours, as CAI can facilitate transactions at any time of day or night. This is very different from traditional online availability, where customers can purchase products at any time, but live customer support is limited to working hours only. Finally, most conversational agents are equipped with high-quality multilingual translators and can interact with customers in many

M. Morzy, *Spoken Language Processing*, Studies in Computational Intelligence 1205,
https://doi.org/10.1007/978-3-031-88566-2_2

languages. Thus, CAI can overcome language barriers, enabling businesses to cater to a global customer base, and at the same time allowing users to interact in their preferred language.

According to the "Global Conversational AI Market" report [9], the market size for CAI is projected to reach $32.62 billion by 2023, growing at a CAGR of 20% in the forecast period covered by the report (2023–2030). The volume of interactions handled by CAI in many industries increased by as much as 250%, with the global COVID-19 pandemic being one of the important accelerators of the adoption of CAI by the public. To illustrate this phenomenon, consider that within just three years the usage of AI for digital marketing skyrocketed from 29% in 2018 to 84% in 2020. As of February 2022, 53% of US adults interacted with a conversational agent for customer service in the last year, and over 3 billion chatbot applications were accessed in that same year. The top three reasons US consumers use conversational agents are business hours (18%), product information (17%), and customer service requests (16%). This wide acceptance of CAI is echoed by 73% of decision-makers who, when surveyed in May 2021, stated that conversational agents will be the most used customer service channel in 2024.

The architecture of CAI, such as voicebots or text-based chatbots, starts with a user's input. For voicebots, this input is processed through Automated Speech Recognition (ASR), which transcribes spoken language into written text. In the case of text input, Natural Language Understanding (NLU) is employed to decipher the user's intent and extract relevant information. Following this, the Dialog Management component comes into play. This part of the system is responsible for determining the appropriate response based on the user's intent and the context of the conversation. It uses complex algorithms, often powered by machine learning and deep neural networks, to maintain the flow of conversation and ensure the responses are contextually appropriate. The final stage involves the Natural Language Generation (NLG) module, which converts the system's response from a machine language into human-readable or audible output.

2.2 Conversational Agents

Conversational AI covers a wide array of digital agents, each with unique capabilities and functionalities. By far, the most popular form of conversational agents are traditional chatbots with the history of the technology going back to the middle of the 20th century. Recent years have witnessed a true explosion of possibilities, including statistical chatbots, voicebots (voice assistants), virtual assistants, and assistants backed by LLMs. Below we present brief descriptions of conversational agents.

- *rule-based chatbots*: these are basic answer and response machines, operating using sets of pre-defined rules, scripts, and exceptions, to generate responses to user inputs. The conversational capabilities of rule-based chatbots are very limited and highly dependent on the keyword search and matching between user inputs and rules. Rule-based chatbots are simple and easy to implement but lack the ability to learn and adapt to new situations. Since they do not utilize NLP or dialogue management, they do not improve over time, which makes their inclusion in the realm of CAI questionable.

- *statistical chatbots*: also known as machine learning-based chatbots, these agents use algorithms and statistical models to analyze user inputs and generate appropriate responses. They can learn and improve over time but require larger amounts of training data and computational resources. By using machine learning algorithms, particularly NLP, to understand user intent and generate responses, statistical chatbots can handle more complex interactions compared to rule-based chatbots.
- *voicebots*: also known as voice assistants or voice-based chatbots, are a form of CAI that interacts with users through spoken language. They utilize advanced speech recognition, text-to-speech, and natural language understanding technologies to interpret voice commands, process them, and deliver audible responses, but their ability to handle context is limited and dialogue state management is rudimentary. Voicebots can perform a wide range of tasks, from answering queries and setting reminders to controlling smart home devices and more. Prominent examples include Amazon's Alexa, Google Assistant, and Apple's Siri.
- *virtual assistants*: these are advanced forms of CAI that can interact with users, understand their commands, and perform complex tasks for them, operating both in audio and text modalities. They combine the capabilities of text-based and voice-based interactions, enabling them to function across various devices and platforms. Apart from answering queries and providing information like a chatbot, virtual assistants can also perform actions such as setting reminders, sending emails, making phone calls, or controlling smart home devices. Two defining features of virtual assistants are the ability to maintain the long-term context of the conversation (e.g. by assisting a customer in the purchase or providing help to an employee executing a complex process) and a sophisticated dialogue management system.
- *LLM-based agents*: the newest iteration of conversational agents is based on Large Language Models. These agents use deep learning to generate human-like text, understanding and generating responses based on context, and providing a more natural, dynamic conversation experience. Each of these agents represents a step forward in the evolution of CAI, offering increasing sophistication and closer approximation to human conversation. Notable examples of such agents include ChatGPT from OpenAI, Bard from Google, Bing Chat from Microsoft, and Claude 3.5 from Anthropic.

2.2.1 *Taxonomy of Chatbots*

Chatbots can be categorized into several main categories based on their functionality, purpose, and interaction style. Customer service chatbots are designed to assist customers with their queries and provide quick answers, commonly found on e-commerce websites and customer service platforms. Conversational chatbots aim to replicate human interaction in natural language, offering a personalized experience on messaging apps and social media platforms. Task-oriented chatbots specialize in performing specific tasks like booking appointments or providing product information, prevalent in sectors such as healthcare, finance, and travel. Entertainment chatbots focus on engaging users through jokes, games, or trivia, often utilized in interactive environments. Educational chatbots deliver educational content, explaining complex concepts and answering questions, frequently used in education and professional development. Personal assistant chatbots help users with daily tasks like scheduling and reminders, commonly featured in virtual assistants. Healthcare chatbots offer healthcare-related information, such as symptom checking and appointment scheduling, heavily used in telemedicine and patient engagement. Financial services chatbots provide banking, investing, or insurance services, typically seen in online banking and investment platforms. Travel chatbots provide travel-related information, including flight schedules and hotel bookings, often employed in travel websites and mobile apps. Lastly, marketing chatbots promote products or services by sending promotional messages and offering discounts, prominently used in e-commerce, retail, and B2B industries. Of course, this list is far from complete. Each category has its own unique features, functionalities, and use cases. As the CAI continues to evolve and improve over time, many new application domains and functionalities are guaranteed to emerge. But functionality is not the only dimension that can be used to characterize chatbots. Another important feature of chatbots is their internal architecture and the scope of models used under the hood to power their conversational abilities. From this point of view, we may distinguish between the following types of chatbots:

- rule-based vs. machine learning-based: as discussed previously, rule-based chatbots are simple programs with pre-defined sets of rules, keyword-based matching of scripts to user inputs, and the lack of dialogue management. Machine learning-based chatbots, on the other hand, use statistical models to generate responses and can somewhat learn over time, albeit to a limited degree.
- text-based vs. voice-based: quite self-explanatory, text-based chatbots communicate through text interfaces such as messaging apps, SMS, or email. Voice-based chatbots use voice recognition technology to understand and respond to spoken commands. They are commonly used in virtual assistants, smart home devices, and voice-controlled gadgets.
- custom vs. off-the-shelf: Custom chatbots are built specifically for a particular organization or industry, tailored to their unique needs and requirements. They offer greater flexibility and customization but require significant development resources and expertise. Off-the-shelf chatbots are pre-built solutions available

in the market, often with pre-defined templates and features. They are easier to deploy and maintain but may not meet specific organizational needs.
- integrated vs. decoupled: Integrated chatbots are part of a larger system or platform, often integrated with other systems such as CRM, ERP, or marketing automation tools. They offer seamless integration and coordination across multiple channels and systems. Decoupled chatbots operate independently, without any integration with other systems. They are simpler to develop and maintain but may not offer the same level of integration and coordination as integrated chatbots.
- human-like vs. non-human-like: Human-like chatbots are designed to mimic human conversation and behavior, using natural language processing and machine learning algorithms to understand and respond to user inputs. They are often used in customer service, tech support, and sales applications. Non-human-like chatbots are designed to provide straightforward, factual information or perform specific tasks, without attempting to mimic human conversation. They are often used in applications such as product recommendation, order tracking, or scheduling appointments.

2.2.2 Chatbots and Virtual Assistants

Oftentimes, the terms "chatbot" and "virtual assistant" are used interchangeably. We have already seen that these two monikers refer to two different types of software artifacts. Nevertheless, it is worth stating these differences explicitly. Four main aspects that can be used to distinguish between chatbots and virtual assistants are: the level of intelligence, the degree to which the agent utilizes natural language processing, the spectrum of tasks the agent is capable of executing, and the underlying technology.

1. Intelligence

    - Chatbots are typically text-based and are programmed to reply to only a certain set of questions or statements. If the question asked falls outside of this set, they will fail. Chatbots cannot hold long human interaction. Traditionally, they are text-based but audio and pictures can also be used for interaction. They provide more like an FAQ type of interaction.
    - Virtual assistants offer a much more sophisticated interaction. They understand not just the language but also the meaning of what the user is saying. They can learn from experience and add some unpredictability to their behavior.

2. Natural language processing

    - Chatbots do not have high language processing skills, they pick certain words from the input and return a programmed response. Chatbots have a structured dialogue and they cannot handle conversations or complex questions that fall outside the scope of the chatbot's area of deployment.

- Virtual assistants employ both natural language processing and natural language understanding. Virtual assistants can understand slang used in everyday conversations, as well as analyze and respond to the sentiment present in the user's input. This allows virtual assistants to have more "natural" conversations.

3. Tasks

- Chatbots have a narrowly defined use and do not have general-purpose algorithms allowing them to perform a wide spectrum of tasks.
- Virtual assistants have a wider scope and can perform a range of tasks. Their usefulness is not limited to a single input or output modality. While still limited, the range of tasks covered by virtual assistants is much broader and usually focused on a particular context (e.g., managing home appliances, hands-free use of phone) rather than specific application domains.

4. Technology

- Traditional chatbots utilize a technology known as *intent discovery* to determine the purpose or goal of a user's input. For instance, when a user types, "What's the weather like today?", the intent could be `[get weather update]`. Once the intent is identified, the bot generates a predefined response. In more complex queries, chatbots employ *slot filling*, a process where they identify and extract necessary pieces of information, or slots, from the user's input to complete a task. For instance, in the query "Book me a flight to Paris on Friday", the bot would identify "Paris" as the `[destination slot]` and "Friday" as the `[date slot]`. In traditional chatbots, the intent discovery and slot-filling processes are often rule-based, requiring predefined patterns and responses, and may not always handle unexpected or complex queries effectively.
- Virtual assistants use deep learning to enable comprehension of context, sentiment, and complex commands. This technology allows virtual assistants to continually learn from their interactions, improving their performance over time. To develop virtual assistants, several popular frameworks and APIs are used, including Google's DialogFlow, Amazon's Lex, Microsoft's LUIS (Language Understanding Intelligent Service), and Rasa AI. These tools provide powerful features for speech recognition, intent discovery, slot filling, and conversation flow management, simplifying the process of building sophisticated virtual assistants.

As we can see, a chatbot and a virtual assistant, while both use CAI and NLP, differ in terms of intelligence, scope of tasks, and underlying technology. A chatbot typically operates within a limited scope, performing specific tasks, often in a rule-based or scripted manner. A virtual assistant is capable of performing a wide array of tasks. It uses advanced NLP, machine learning, and often deep learning, to understand, learn, and respond to user queries in a more conversational and personalized manner.

Virtual assistants often have the ability to integrate with multiple platforms and systems, providing a more comprehensive service.

2.3 History of Chatbots

Alan Turing, the pioneer in the field of computing and artificial intelligence, made a consequential claim in his famous "Computing Machinery and Intelligence" paper [23]. In this seminal work, Turing proposed what is now known as the Turing Test, a measure of a machine's ability to exhibit intelligent behavior equivalent to, or indistinguishable from, that of a human. Turing hypothesized that a machine would eventually be constructed that could engage in natural conversation with humans, mirroring human-like responses. This was a bold assertion at the time, given the relatively primitive state of computing technology. Nevertheless, Turing's visionary outlook has influenced the development of AI, particularly in the realm of natural language processing and conversational AI.

Commonly considered to be the first chatbot capable of human-like conversation, ELIZA was developed in the 1960s by Joseph Weizenbaum at MIT's Artificial Intelligence Laboratory [27]. It was designed to simulate a conversation with a psychotherapist, and it used a simple pattern-matching algorithm to generate responses based on the user's input. ELIZA implemented the concepts of person-centered therapy (PCT) developed originally by psychologist Carl Rogers [17]. According to the principles of PCT, patient statements are to be reflected back to the patient in the form of questions and prompts in order to encourage a more in-depth exploration of emotions and thoughts. ELIZA's pattern-matching algorithm worked by analyzing the user's input and comparing it to a set of pre-defined patterns or keywords. If a match was found, ELIZA would respond with a pre-written response that was associated with that particular pattern. Most patterns revolved around turning input statements into paraphrases or questions, with linguistic marks of empathy and understanding embedded in responses. The algorithm was relatively simple and did not take into account the context or nuances of the conversation. ELIZA's responses were also limited by the scope of her training data, which consisted of a small number of conversations that had been manually transcribed from therapy sessions. As a result, ELIZA's responses were often generic and did not address specific issues or concerns raised by the user. Despite its limitations, ELIZA was a groundbreaking achievement in the field of natural language processing and paved the way for the development of more advanced chatbots in the future.

The next chatbot to capture popular attention was Parry, an innovative early chatbot developed in 1970 by psychiatrist Kenneth Colby at Stanford University [5]. The chatbot was designed to simulate a person with paranoid schizophrenia and exhibit behaviors associated with this mental disorder, such as exhibiting suspicion, harboring resentments, and demonstrating a tendency toward grandiose thinking. Parry was unique in its ability to engage in conversation with human users, often leading to immersive and complex interactions.

One of the first attempts to bring chatbot technology to a wide audience was Dr. Sbaitso. It was developed in the early 1990s by Creative Labs as a part of the Sound Blaster technology. Presented as an artificial intelligence psychologist, Dr. Sbaitso was designed to interact with users through text-based conversations. The chatbot was programmed to analyze and respond to the user's input, simulating a psychotherapist's role. One of the notable features was its text-to-speech function, which enabled the software to verbalize the text responses using a synthesized voice, a novel feature for its time. The main drawback of Dr. Sbaitso was a very limited catalog of responses, with many conversations ending with sentences such as "It is not my problem!" While its conversational capabilities were relatively rudimentary by today's standards, Dr. Sbaitso represented an early effort to create interactive, conversational AI and showcased the potential for computers to engage in more human-like interactions.

The next notable development in the history of chatbots was A.L.I.C.E. or Artificial Linguistic Internet Computer Entity. It was developed by Richard Wallace in the mid-1990s [26]. A.L.I.C.E. utilized a pattern matching and heuristic processing approach to carry out conversations with users. It was designed to simulate human conversation through the use of an XML schema known as AIML (Artificial Intelligence Markup Language). A.L.I.C.E. was capable of engaging in meaningful interactions and was instrumental in furthering the development of conversational agents. The chatbot gained recognition for its sophistication and conversational abilities, winning the Loebner Prize, an annual competition in artificial intelligence that judges the "most human-like" chatbot, three times in 2000, 2001, and 2004.

The beginning of the 21st century witnessed an explosion in the popularity of instant messaging. Hundreds of applications were developed to allow human communication over the Web. As online connectivity became widespread, it provided a vast platform for the implementation and testing of AI technologies. This increased accessibility to the internet, coupled with advancements in natural language processing and machine learning, facilitated the rapid evolution and sophistication of chatbots. SmarterChild [13] was a pioneering chatbot developed by ActiveBuddy and launched on AOL Instant Messenger and MSN Messenger in the early 2000s. SmarterChild was designed to provide a range of services including weather updates, news, and movie show times, among other things, all through the simplicity of instant messaging. Users could ask questions or request information on a wide range of topics using text-based conversations, and SmarterChild would generate answers based on its knowledge base. The key aspect of SmarterChild, distinguishing this chatbot from previous products, was its versatility. SmarterChild could converse about sports scores, stock market updates, trivia games, weather forecasts, and many more. For the first time users had access to a chatbot that removed the necessity to visit multiple websites and instead presented aggregated information in conversational form. SmarterChild could not understand truly the context of conversations, nor could it maintain a stateful dialog, but it was still regarded as a significant leap in the integration of AI into everyday digital communication.

As the domain of AI advanced and matured, so did the technology powering chatbots. The first truly advanced chatbot was Cleverbot[1] developed by R. Carpenter. Unlike traditional chatbots that rely on pre-programmed responses, Cleverbot utilized a vast database of prior conversations filtered through fuzzy logic to generate its responses, effectively "learning" from its interactions with users. This method allowed Cleverbot to generate remarkably human-like dialogue, capable of passing the Turing Test under certain conditions.

The development of chatbot technology is best followed by analyzing the winners of the Loebner Prize for the most human-like chatbot. In 2013 and then from 2016 to 2019 the prize went to S. Worswick from Pandorabots for his work on Mitsuku (later renamed to Kuki AI).[2] This AI-powered chatbot boasts the ability to engage in detailed and contextual conversations with users, simulating a remarkably human-like interaction. Built on AIML technology, Kuki AI learns from every conversation it has, thus constantly improving its responses. Its ability to understand context, handle an extensive array of topics, and learn from its interactions sets it apart from many other chatbots. The recognition received by Kuki AI (it served as an inspiration for the movie "Her" by Spike Jonze) highlights the advancements made in the field of AI and its potential to create increasingly sophisticated conversational agents. The chatbot has been included in Facebook Messenger, Twitch group chat, Telegram, Discord, and Skype.

Following the success of Kuki AI, many similar virtual personas have been created. One notable example is Rose,[3] a chatbot developed by B. Wilcox. Known for her sophisticated conversational abilities, Rose is designed to simulate a 31-year-old woman with an intriguing background–she is a security analyst and a hacker. This detailed backstory enhances the depth and realism of her interactions. Rose made her mark in the field of AI by winning the Loebner Prize in 2014 and 2015. Rose uses a unique approach to generate her responses, rather than relying exclusively on machine learning, she utilizes a handcrafted knowledge base and conversation flow. This allows for a high degree of control over her dialogue, contributing to her ability to engage in realistic, contextually accurate conversations.

Another example is Xiaoice,[4] an advanced chatbot technology developed by Microsoft. Launched in 2014, Xiaoice, which translates to "Little Ice" in English, is designed to simulate human-like text conversations and has gained considerable popularity, especially in China. Xiaoice is built to offer more empathetic and interactive dialogues, it utilizes sentiment analysis, natural language processing, and adaptive learning capabilities to understand and respond to user emotions effectively. Xiaoice's unique ability to remember past conversations and context allows it to engage in more personalized and long-term interactions with users. In particular, Xiaoice uses Markov Decision Process which optimizes the service for long-term user engagement. To make interactions even more human-like, Xiaoice uses extensively

[1] https://www.cleverbot.com.

[2] https://www.kuki.ai.

[3] https://rose.ai.

[4] https://www.xiaoice.com.

both intelligence quotient and emotional quotient to build emotional connections with users and satisfy their need for social belonging.

Kuki AI and Xiaoice both represent the transition from chatbots to virtual assistants. Without a doubt, the same can be said about Melody, a specialized medical chatbot developed by Baidu Research. Introduced in 2016, Melody is designed to assist doctors by providing information related to medical inquiries. The chatbot uses deep learning algorithms and natural language processing to understand medical terminology and context. It can extract pertinent details from a patient's symptoms description and suggest likely diagnoses. In addition, Melody can interact with patients, ask follow-up questions to gather more information, and even provide general medical information. Similar functionality has been implemented in other chatbots, such as HealthTap, Koko, and Sensely.

Tay, an ambitious chatbot project launched by Microsoft in 2016, stands as a spectacular case study in the precarious nature of AI learning [29]. The concept was innovative yet simple: Tay would learn, grow, and evolve directly from interactions with Twitter users, progressively enhancing its conversational abilities. However, this seemingly brilliant idea crumbled within a mere 24 hours of its implementation. The unchecked influx of toxic language from users, rife with racism, xenophobia, and fascism, transformed Tay from an innocent chatbot to a disturbing echo of the internet's darkest corners. This swift corruption forced Microsoft to unplug Tay, leaving a stark reminder of the potential pitfalls in AI development when sufficient safeguards are not in place.

The end of the second decade of the XXI century witnessed the explosion of large language models. One of the most anticipated models was Generative Pre-trained Transformer 2 (GPT-2) released by OpenAI in November 2019. It was trained on the corpus of 7000 books and 8 million webpages, a ten-fold increase in the training dataset size in comparison with its predecessor, GPT model. DialoGPT is a variant of the GPT-2 model, specifically fine-tuned for generating conversational responses [30]. The ability of DialoGPT to produce coherent and human-like sounding dialogues was the result of fine-tuning on a massive dataset of 147 million multi-turn dialogues from Reddit discussion threads. In addition, DialoGPT uses a different training objective to its base GPT-2 model. The training objective of DialoGPT focuses on maximizing the likelihood of the next response given the context of the conversation. This results in responses that are more likely to be natural and engaging. DialoGPT has been shown to achieve state-of-the-art performance on a variety of dialogue generation benchmarks. Since its release, DialoGPT has been used to build a variety of CAI systems, including chatbots, virtual assistants, and educational tools.

The quickly rising popularity of language models released by OpenAI forced Google to enter the market. Meena [1] was the first neural conversational AI model developed and released by Google. Meena's architecture is based on a seq2seq model and a variant of the Transformer. It comprises 2.6 billion parameters, making it one of the largest language models at the time of its release. The model was trained on a massive 341 GB dataset of public web text, including 28 GB of dialogue data from Reddit conversations. One of the distinctive features of Meena is its use of the SSI

(Sensibleness and Specificity Average), a novel metric to evaluate the performance of chatbots. This metric measures how sensible and specific a chatbot's responses are, ensuring higher-quality dialogues. Meena represents a significant stride toward achieving truly conversational AI, and the model has been praised for its natural and sensible conversations with humans.

No chatbot has managed to capture the public's attention to the degree of ChatGPT, the direct descendant of the DialoGPT. It is based on the GPT-3.5 and GPT-4 foundation models, which were fine-tuned to target conversational usage. ChatGPT uses the decoder-only Transformer architecture, and the defining feature of the model is its use of Reinforcement Learning with Human Feedback (RLHF) [20]. RLHF is a method used in machine learning where an AI model learns to make decisions based on feedback it receives from human interactions. In this approach, the model is initially trained on a dataset where humans demonstrate the task. Then, a reward model is created where humans rank different model-generated outputs based on their quality. The AI model is then refined through reinforcement learning, where it learns to optimize its actions based on the reward feedback. This process of learning and refining continues iteratively, with the model improving its performance over time. It is essentially a way of teaching AI to replicate and improve upon human decision-making processes. The success of RLHF was made possible by the development of Proximal Policy Optimization (PPO), a new algorithm to discover the most efficient policy in reinforcement learning [6]. In the context of reinforcement learning, the term *policy* refers to the model's behavior and to the choice of action in response to a given input. PPO is designed to optimize this policy in a more sample-efficient and stable way compared to other methods. It does so by limiting the changes to the policy in each update, ensuring the new policy does not deviate too far from the old one. This proximity constraint makes the learning process more stable and prevents harmful fluctuations in the model's performance. The use of RLHF and the introduction of the new policy discovery algorithm allowed ChatGPT to attain unprecedented quality of conversations.

Unfortunately, this increased efficiency came at the expense of vulnerable communities. As reported by the "Time" magazine [16], OpenAI outsourced the task of RLHF to Sama Group, a training-data company based in San Francisco, California. This company was using low-paid data labellers in Kenya to build a detector of toxic content built into ChatGPT. The task amounted to manually labeling textual descriptions of sexual abuse, hate speech, and violence. The labellers were exposed to images and descriptions of bestiality, rape, sexual slavery, and graphic detail of death, violence, or serious physical injury, for a take-home wage of between \$1.32 and \$2 per hour, depending on the contract. One of the employees described the assignment as "torture", citing an example of a labeling task including the graphical description of an act of bestiality in front of a child. The story described by "Time" unveiled a disturbing and shameful aspect of the modern AI industry, which depends on the exploitation of hidden human labor in the Global South.

One of the last products to enter the intelligent chatbot arena is BlenderBot [19]. BlenderBot is a state-of-the-art chatbot developed by Meta AI. Its latest version released in August 2022 is based on Meta's OPT-175B language model, and the name reflects the chatbot's ability to merge multiple conversational skills at once. Similarly to other chatbots, BlenderBot is built on the Transformer architecture and trained using RLHF. BlenderBot boasts several features that set it apart from previous chatbots. It has the ability to engage in open-domain conversations [28], meaning it can carry on a conversation about virtually any topic. It can also demonstrate empathy, express knowledge, and exhibit a distinct personality, providing a more human-like interaction. Key technical developments that allow these advancements are the search-augmented language model and the combination of the generator-classifier approach to language modeling. BlenderBot uses SeeKeR [18], the search-augmented language model that augments token generation by iteratively calling a Transformer module to generate a web search query, build a knowledge sequence, and generate the final token. By querying external knowledge sources, such as Wikipedia or Google, BlenderBot is not limited to knowledge memorized in its parameters, but can also discuss themes that were not contained in its training data. In addition, the presence of external knowledge helps the model to limit the hallucination and stay topical during long conversations. The second development is the use of a new architecture for output token generation. BlenderBot uses DIRECTOR [2], a simple architecture where each token generated in the output sequence is additionally classified as either desirable or undesirable. The training of the model uses joint optimization of the standard language modeling data and data labeled with desirable and undesirable sequences. This feature protects BlenderBot from producing toxic output. When BlenderBot was released to the public and the first batch of 70 000 conversations were collected, only 0.11% of responses were flagged as inappropriate, 1.36% as nonsensical, and 1% as off-topic. Given the fact that conversations were open-domain and conducted outside of the safety sandbox, this is an impressive result. As an additional consequence, analyzing the results of public open-domain conversations with the BlenderBot led to the development of SafetyMix [7], a benchmark to measure methods for detecting toxic language in human conversations in a variety of adversarial settings. In addition, the researchers compiled SaFeRDialogues [24], a task and dataset of 8 000 graceful responses to conversational feedback about safety failures. Initial experiments have shown that fine-tuning of a chatbot on SaFeRDialogues produces conversations that are rated as more civil by human annotators.

In closing, the field of intelligent conversational chatbots is marked by fierce competition and relentless pursuit of innovation and sophistication. The significant financial incentives have motivated a host of powerful players, each vying to push the boundaries of what is possible. Companies like Google, Meta, and OpenAI have poured substantial resources into exploring the potential of chatbots, resulting in many advancements in AI and NLP. Looking ahead, the future of chatbots looks promising. We can anticipate chatbots becoming increasingly elaborate, capable of understanding and responding to human language with even greater accuracy and nuance. Developments may also include more personalization and adaptability, with chatbots learning and evolving to better meet individual user needs. The integration of chatbots across various sectors, from customer service to mental health, is likely to become even more pervasive. While the technology has come a long way since the times of ELIZA, it is clear that the journey of chatbots is still very much in progress, with many exciting chapters yet to be written.

2.4 History of Voicebots

The evolution of voicebots and voice assistants follows the history of communication technology at the intersection of artificial intelligence, speech recognition, and human-computer interaction. We can trace the beginnings of this technology to the early 1960s, with the advent of systems like IBM's Shoebox, which could recognize spoken digits and a handful of words. Many significant developments in voicebot technology took place over the years, but the true acceleration happened in the last decade, when advancements in AI and machine learning made it possible to develop sophisticated voice-activated systems. These systems, such as Apple's Siri, Amazon's Alexa, Google's Assistant, and Microsoft's Cortana, revolutionized the way humans interact with technology, causing a shift from text-based to voice-based communication. In this section we look at how voicebots and voice assistants have evolved over the years, transforming from a technological novelty to an integral part of our daily lives.

The history of voicebots begins with IBM Shoebox, a landmark in the history of voice recognition technology. Introduced in 1961, this device was one of the earliest attempts to facilitate human-computer interaction through voice commands. The Shoebox was named after its distinctive shoebox-shaped design and was approximately the size of a small suitcase. It contained a microphone, a speaker, and a transistorized computer. The Shoebox was capable of recognizing 16 spoken words, including the digits from 0 to 9, as well as the words "plus," "minus," "total," and "false." Users could perform mathematical operations such as addition, subtraction, multiplication, and division using voice commands. The Shoebox would convert spoken sounds into electrical impulses, which would then be translated into binary code for the machine to comprehend and respond to. While rudimentary by today's standards, IBM's Shoebox was a significant development in the field of speech recognition. Although the Shoebox was never commercialized, it demonstrated that it was possible to build a machine that could understand and respond to human speech, and it inspired a new generation of researchers to work on developing more sophisticated voice-activated devices.

The next milestone in the development of voicebots was the Harpy speech recognition system [8]. The Harpy was developed in the mid-1970s at Carnegie Mellon University. Harpy was designed to recognize approximately 1 000 words, a capacity that was almost revolutionary at the time. The key to Harpy's success was the employment of the dynamic programming algorithm that allowed it to compare an incoming speech signal with all its known models simultaneously. This made Harpy one of the first voice recognition systems to use connected speech, allowing it to understand naturally spoken sentences. In terms of accuracy, Harpy achieved an impressive 90-95% recognition rate, outperforming many of its contemporaries.

Further enhancements to Harpy included the application of Hidden Markov Models (HMM) to voice recognition. HMMs are statistical models that can be used to represent the sequence of acoustic features that are produced when a person speaks [14]. The Harpy system used a large database of speech data to train an HMM for each word in its vocabulary. Harpy would extract a sequence of acoustic features from the user's speech. The system would then use the HMMs in its vocabulary to find the sequence of words that were most likely to have produced the observed acoustic features. Harpy was used in a variety of applications, including dictation, command and control, and voice-activated interfaces.

The first consumer-facing voice recognition systems was Dragon Dictate [11], introduced in the early 1990s by Dragon Systems, and later purchased by Nuance Communications. Dragon Dictate was designed to convert spoken words into written text, enabling users to dictate text directly to a computer rather than typing it. The software used a variety of techniques to recognize speech, including noise cancellation and acoustic modeling. Dragon Dictate also had a built-in vocabulary that could be customized to include specialized terms or jargon. The original system required the user to speak slowly and distinctly, pausing between words. However, despite these limitations, it was a groundbreaking development that opened up new possibilities for people with physical disabilities who found it difficult to use a keyboard or a mouse. Over time, Dragon Dictate evolved into Dragon NaturallySpeaking, which became capable of transcribing speech in real time and recognizing continuous, natural speech. Dragon Dictate opened the door to other commercial products in the space of speech recognition, such as Julie Doll, the first toy that could be taught to react to an owner's voice, or IBM's MedSpeak/Radiology, the world's first continuous-speech recognition dictation product where radiologists would dictate the examination of a patient's X-ray.

The presentation of the history of voicebots would be incomplete without mentioning the infamous Clippy, an interactive office assistant that made its debut in Microsoft Office 97. Although primarily known for its text-based interactions, Clippy was also integrated with speech recognition capabilities in later versions of Microsoft Office. This allowed users to interact with Clippy using voice commands, making it a primitive form of the voice-activated assistants we see today. Clippy was designed to help users navigate the Office suite and perform tasks more efficiently by offering helpful suggestions and tutorials. Despite its original intention to improve user experience, Clippy was often criticized for its intrusive nature and was eventually discontinued in 2007.

A truly groundbreaking moment and the transition from voicebots to voice assistants happened in October 2011. Voicebots are simple programs that can recognize and respond to a limited set of spoken commands. Voice assistants, on the other hand, are more sophisticated programs that can understand and respond to a wide range of natural language commands. Apple's Siri, introduced in 2011 as a feature of the iPhone 4S, represented a significant shift in the field of voice recognition technology. Siri was not merely a tool for transcribing speech to text or executing voice commands, it was a sophisticated AI-powered assistant capable of understanding natural language, context, and intents. Siri could perform a wide range of tasks, from setting reminders and sending messages to answering questions and providing recommendations. This leap in functionality was not the only thing that set Siri apart. As an integrated feature of the iPhone, Siri was one of the first voice assistants to achieve widespread commercial and consumer exposure. Its introduction played a key role in bringing voice recognition technology to the masses, normalizing the use of voice assistants in everyday life. The commercial success and popularity of Siri demonstrated the market's readiness for such technology and paved the way for the introduction of other voice assistants like Amazon's Alexa and Google Assistant.

The huge success of Siri prompted other tech giants to unveil their products as well. Google Now, launched in May 2012, was Google's intelligent personal assistant, designed to provide information and perform tasks for users based on their habits and preferences. Examples of such complex tasks included the use of user's geolocation data (providing information about nearby businesses, traffic conditions, and weather), search history (providing personalized recommendations for news, restaurants, and other products and services), or calendar and contacts data (providing reminders about upcoming events and birthdays). The service was initially available within the Google Search mobile application for Android and iOS, and later integrated into Google's Chrome web browser and many other Google products. Google Now used a natural language user interface to answer questions, make recommendations, and perform actions by delegating requests to a set of web services. It was known for its predictive capabilities, as it could provide information to the user based on their previous behavior. Google Now was eventually replaced by Google Assistant in 2016, which retained many of Google Now's features but added a conversational interface and expanded capabilities.

Obviously, Microsoft had to react to these developments with its own proposal. Cortana, named after the artificial intelligence character in Microsoft's Halo video game franchise, was launched by Microsoft in April 2014. Initially designed for Windows Phone 8.1, Cortana was later incorporated into Windows 10, Xbox One, and various other Microsoft platforms. Cortana used NLP and machine learning algorithms to understand voice commands, answer questions, set reminders, and perform a wide range of other tasks. Cortana's functionality was significantly boosted by its seamless integration with Microsoft's Bing search engine, allowing it to pull information from the web quickly. It also offered predictive capabilities, learning from user's habits and preferences over time to provide personalized assistance. In recent years, Microsoft has shifted Cortana's focus more towards productivity features, integrating it tightly with Microsoft 365 suite of applications.

In November 2014 the fourth tech giant joined the fold. Amazon Echo, commonly known as Alexa, is a virtual assistant AI technology developed by Amazon, first used in the Amazon Echo and the Amazon Echo Dot smart speakers. Alexa is capable of voice interaction, music playback, making to-do lists, setting alarms, streaming podcasts, playing audiobooks, and providing real-time information such as weather, traffic, sports, and news. It can also control several smart devices, acting as a home automation system. Alexa's capabilities are expanded through "skills"–third-party developed capabilities that add to the assistant's functionality. These skills cover a wide range of areas, from ordering a pizza or ride-sharing service to controlling smart home devices. Over time, Amazon has added new features and updated its AI technology, making Alexa more conversational and context-aware. By integrating with Amazon's vast ecosystem of services, Alexa has become a popular choice for home automation and personal assistance tasks.

The success of Siri, Google Assistant, Cortana, and Alexa, prompted the release of hundreds of AI-enabled devices that could work with voice assistants. Google Home, Genie X1 Smart Speaker from Alibaba, Dragon Medical Virtual Assistant from Nuance, Sonos One speakers, smart speakers from JBL, Bose, LG, and Lenovo, Galaxy Home from Samsung, Djingo from Samsung, AirPods from Apple, Samsung Galaxy Buds, Magenta from Deutsche Bank, are just a few examples of devices that entered the market between 2015 and 2020. Another domain where voice assistants became ubiquitous is motorization. Voice assistants have become an integral part of modern vehicles, enhancing the driving experience by facilitating hands-free control of various in-car functionalities. They allow drivers to control navigation, play music, send and receive messages, make calls, and even check the weather, all through simple voice commands. This minimizes distraction, contributing to safer driving. Some car manufacturers have also developed their own voice assistants, such as Mercedes' MBUX and BMW's Intelligent Personal Assistant. These in-car voice assistants can learn and adapt to the driver's habits and preferences. With the advent of connected car technology and IoT, these voice assistants can also interact with smart devices at home, enabling drivers to control aspects like home lighting or heating from their vehicles.

Between 2020 and 2024, significant strides were made in the field of voice assistants. They became more adept at understanding context, dialects, and accents, providing more accurate and personalized responses. The integration of voice assistants with other technological domains also expanded, particularly in the areas of home automation, automotive infotainment systems, and healthcare. In 2021, Google released a new version of its Assistant that is able to understand and respond to over 100 languages. In 2022, Amazon released a new version of its Alexa assistant that is able to control over 100 000 smart home devices. And in 2023, Microsoft released a new version of its Cortana assistant that is able to translate languages in real-time. Innovations like Alexa's Care Hub feature, which provided remote care for the elderly, demonstrated the potential of voice assistants in the healthcare sector. The focus on privacy and security also intensified, with tech companies introducing new features and protocols to protect user data and offer more transparency. Furthermore, the proliferation of 5G technology contributed to faster and more seamless interactions with voice assistants. The trend of creating proprietary voice assistants also gained momentum, with more companies developing their own voice AI systems.

In the future, voice assistants are expected to become even more intuitive and personalized, capable of understanding and responding to users' emotions and moods. They may interact in a more human-like manner and improve conversational capabilities, due to better integration with LLMs. The integration with augmented and virtual reality could create immersive experiences, transforming how we interact with digital content. The fierce competition to remove the limitations imposed by flat two-dimensional screens and move human-computer interaction to the digital realm of virtual or augmented reality drives the research into better than "fingers on a keyboard" modes of communication. If we are successful in ditching flat screens, voice communication (along with body motion) clearly emerges as the best way to interact with computers. On the other dimension, advancements in AI and data analytics could enable voice assistants to provide proactive assistance, predicting users' needs based on historical data and patterns. For instance, they could automatically schedule meetings, remind users of important events, or suggest personalized health and wellness advice. However, these developments would also require robust privacy and security measures to ensure that the personal data used to enhance the functionality of voice assistants is adequately protected.

2.5 Conversational LLMs

A chapter on Conversational AI would certainly be incomplete without a section exploring the recent advancements in LLMs, specifically those fine-tuned for conversation via reinforcement learning from human feedback (RLHF). These models significantly changed the landscape of CAI. They can generate more natural, contextually accurate, and engaging responses, bringing us closer to the goal of seamless human-machine communication. RLHF serves as a key technique in training these models, enabling them to adapt and learn from real-world interactions, improving their performance over time. With their ability to understand and generate human-like text, these models have improved the quality of AI-driven conversations and opened up new possibilities in diverse applications such as virtual assistants, customer service bots, and more. Ignoring these developments would mean overlooking some of the most transformative progress in the field.

The introduction of the Transformer architecture [25] sparked the explosion of training of very large language models. The training corpora became so big, that the models first got the moniker *large*, and then *foundational*. A foundational large language model is a type of language model that has been trained on a broad range of internet text. These models have the capability to generate human-like text based on the input they receive. They are called "foundational" because they can be used as a base for a wide variety of applications, from drafting emails and writing code to creating written content, answering questions, tutoring in multiple subjects, translating languages, simulating characters for video games, and much more. The rise of these models has been fueled by the exponential increase in computational power, the availability of vast amounts of data, and significant advancements in training techniques. From generating coherent long-form content to translating languages with remarkable accuracy, these models have set the stage for a new era of AI applications. Usually, the term "foundational" is reserved for the largest models, such as GPT-4, PaLM, LlaMA, or Falcon.

The conversational nature of generative models has been always present, encoded in the vastness of their parameter space, but it has never been the primary objective of model training. When the quality of the generated text started to approach the quality of the average human's writing, it became obvious that these models could be used for conversations. However, if traditional language models were to enter human conversations, at least three major challenges have to be addressed: understanding the flow of conversation, maintaining a long context of conversation, and focusing on the topic of conversation.

Human conversations have a very different structure from the traditional written text used to train LLMs. A conversational model has to recognize the interplay between statements and responses, as well as identify cues for when to initiate or conclude a thread of conversation. Traditional LLMs can produce very coherent text even if their context window (the number of tokens they can ingest as the input) is relatively low because the generation of text usually depends on tokens that are in close proximity. For a smooth conversation, a model has to maintain a "mental state" of the conversation, which often involves referring to information mentioned earlier in the conversation. Conversational AI models often struggle with longer conversations. As dialogues extend, the models must retain relevant information, recall earlier parts of the conversation, and ensure continuity in responses. This is challenging due to the limited memory of AI models and the complexity of managing and prioritizing information over extended dialogues. Finally, generative language models have the tendency to veer off the initial topic as the generation of output continues. An effective conversation often requires staying on topic, which can be difficult for AI models. They can sometimes generate responses that, while grammatically correct and plausible, can diverge from the original topic of conversation. Ensuring that the AI model stays on track and responds in a contextually appropriate manner is a significant challenge in training conversational AI models.

The solution to the aforementioned challenges proved to be the application of specific fine-tuning processes. The original training of Transformer models involved predicting masked words, predicting the next sentence, etc. The first obvious choice was to perform additional model fine-tuning with a specific corpus of dialogue data. This allowed the models to refine and adapt their text generation capabilities to the context of human conversation, learning to produce responses that are contextually appropriate and coherent. The second choice was to reinforce the desired responses of a model as judged by human annotators. This technique is known as reinforcement learning from human feedback (RLHF), where the model gradually improves its conversational skills based on feedback from human interactions [31]. We have already presented this technique when discussing the development of DialoGPT, and its successor, ChatGPT.

The introduction of ChatGPT was a true technological earthquake. The model instantly conquered the front pages of newspapers all around the world and stormed prime-time television programs. Suddenly, an AI model was accessible to all via a simple Web-based interface and ready to generate jokes, memes, homework assignments, essays, reviews, and conversations. Immediately other companies rushed to present their own conversational models that could compete with ChatGPT.

The most anticipated model in the realm of conversational AI was Google's Bard (Bidirectional and Auto-Regressive Transformers for Decoding) [12]. Bard was initially based on the LaMDA large language model, but it subsequently switched to

PaLM as the underlying language model. Bard has been in development since 2021, but little has been released to the public. When OpenAI unveiled ChatGPT in November 2022, Google was forced to accelerate research which led to a limited publication of Bard in February 2023. Since the hype around generative AI and conversational AI had been mostly hijacked by OpenAI by that time, the initial response to Bard was far from enthusiastic. Bard is designed to improve multi-turn conversation understanding and generation. Unlike traditional models, Bard utilizes a unique approach where it encodes all previous dialogue history and decodes future responses in a single pass. This allows it to handle complex, multi-turn conversations with greater accuracy and fluency. The model uses the strengths of both Transformer and RNN architectures, combining their capabilities to efficiently manage the dependencies between different turns in a conversation. Despite receiving mixed reviews upon its release, Bard continues to be integrated into Google's products and tries to compete with ChatGTP and other similar models.

Another interesting proposal released in response to ChatGPT is the Claude family of models developed by Anthropic. While specific details about Claude's development timeline and unique features are not publicly available, the general aim of Anthropic in creating models like Claude is to progress the field of AI toward models that users can understand and effectively control, an approach called the Constitutional AI [3]. This includes placing strong emphasis on transparency, robustness, and the ability to provide detailed insights into the model's decision-making processes. By creating models that align with human values and can be directed and corrected by human users, Anthropic seeks to make significant advances in the safety and utility of AI. Also, Claude 2 was the first model to break the 100 000 token context limit. The ability to maintain such a huge context is regarded as a safeguard against hallucination and topical drift in conversations.

The LLaMA (Large Language Model Meta AI) family of models [10], developed by Meta AI, represents a significant advancement in the field of conversational LLMs. The first iteration, LLaMA 1, was introduced as a series of models with varying sizes, designed to balance performance and computational demands. Following this, LLaMA 2 [22] expanded on its predecessor with improvements in architecture and training methods. The latest version, LLaMA 3, has been trained on dataset comprising over 15 trillion tokens sourced from publicly available online data. Meta applied thorough data curation methods, such as heuristic and NSFW filters, semantic deduplication, and text classifiers, to ensure high-quality input for training the LLaMA models. About 5% of this training data comprises multilingual content in over 30 languages, aiming to enhance the model's utility across different linguistic contexts, though it is expected to perform best in English. The 8 billion parameter version of the model has a knowledge cutoff in March 2023, while the 70 billion version extends to December 2023, allowing the model to incorporate relatively recent information. LLaMa is an open-weight model. Being an open-weight model means that the parameters or "weights" of the LLaMA models are publicly available, allowing researchers and developers to access, modify, and build upon the models without restriction. This openness has sparked a revolution in the AI community, promoting collaboration and innovation in developing open-weight and open-source models.

As a result, several models have been built upon the LLaMA foundation, such as Alpaca [21] and Vicuna [4], which use LLaMA as its foundation.

Finally, one should mention the alternative direction of foundational model fine-tuning which is gaining momentum: the paradigm of instruction following. Instruction-following large language models are a type of LLMs that are specifically designed to follow instructions and complete requests thoughtfully. They are trained on a massive dataset of text and code, which includes examples of human dialogue and instructions. This training allows them to learn how to understand and follow complex instructions, even if they are open-ended, challenging, or strange. Instruction-following LLMs have a wide range of potential applications. They can be used to develop chatbots that provide customer service, education, entertainment and even can be used to develop new types of creative writing tools. A prominent example of this family of models is InstructGPT [15]. This model is a variant of GPT-3 designed to follow instructions provided in a prompt. The model was trained using a similar method to ChatGPT, but with an important difference: instead of conversation data, it was trained on datasets transformed into an instructional format. The model aims to understand and execute a wide range of complex instructions within its text inputs. Unlike CAI, which focuses on generating human-like dialogue, InstructGPT focuses more on understanding and executing instructions provided in its prompts.

In conclusion, the realm of CAI has experienced significant advancements over the last years, transforming the way we interact with technology. From simple rule-based chatbots to sophisticated voicebots, the evolution has been remarkable. Virtual assistants like Siri, Google Assistant, and Alexa have become household names, changing how we search for information, manage our schedules, and control smart home devices. More recently, the development of highly advanced conversational language models, such as Google's Bard model or OpenAI's ChatGPT, has further pushed the boundaries of what CAI can achieve. Looking ahead, the continual refinement and development of these technologies promise an era where AI will not only understand and respond to our queries but will also anticipate our needs and provide proactive assistance.

References

1. D. Adiwardana, M.-T. Luong, D.R. So, J. Hall, N. Fiedel, R. Thoppilan, Z. Yang, A. Kulshreshtha, G. Nemade, Y. Lu, et al., Towards a human-like open-domain chatbot (2020). arXiv preprint arXiv:2001.09977
2. K. Arora, K. Shuster, S. Sukhbaatar, J. Weston, Director: Generator-classifiers for supervised language modeling (2022). arXiv preprint arXiv:2206.07694
3. Y. Bai, S. Kadavath, S. Kundu, A. Askell, J. Kernion, A. Jones, A. Chen, A. Goldie, A. Mirhoseini, C. McKinnon, et al., Constitutional ai: Harmlessness from ai feedback (2022). arXiv preprint arXiv:2212.08073
4. W.-L. Chiang, Z. Li, Z. Lin, Y. Sheng, Z. Wu, H. Zhang, L. Zheng, S. Zhuang, Y. Zhuang, J.E. Gonzalez, I. Stoica, E.P. Xing, Vicuna: An open-source chatbot impressing gpt-4 with 90%* chatgpt quality (2023)

5. K. Mark Colby, *Artificial Paranoia: A Computer Simulation of Paranoid Processes*, vol. 49 (Elsevier, 2013)
6. L. Gao, J. Schulman, J. Hilton, Scaling laws for reward model overoptimization, in *International Conference on Machine Learning* (PMLR, 2023), pp. 10835–10866
7. D. Ju, J. Xu, Y.-L. Boureau, J. Weston, Learning from data in the mixed adversarial non-adversarial case: Finding the helpers and ignoring the trolls (2022). arXiv preprint arXiv:2208.03295
8. B.T. Lowerre, The harpy speech recognition system (1976)
9. SkyQuest Technology Consulting Pvt. Ltd., Global conversational ai market (2023). https://www.skyquestt.com/report/conversational-ai-market. Accessed 21 Sep 2023
10. L. Van Der Maaten et al., The llama 3 herd of models (2024). arXiv preprint arXiv:2407.21783. Accessed 9 Sep 2024
11. M.A. Mandel, A commercial large-vocabulary discrete speech recognition system: Dragondictate. Lang. Speech **35**(1–2), 237–246 (1992)
12. J. Manyika, *An Overview of Bard: An Early Experiment with Generative AI* (Tech. rep, Technical report, Google AI, Technical report, 2023)
13. G. Molnár, Z. Szüts, The role of chatbots in formal education, in *2018 IEEE 16th International Symposium on Intelligent Systems and Informatics (SISY)* (IEEE, 2018), pp. 000197–000202
14. B. Mor, S. Garhwal, A. Kumar, A systematic review of hidden markov models and their applications. Arch. Comput. Methods Eng. **28**, 1429–1448 (2021)
15. L. Ouyang, W. Jeffrey, X. Jiang, D. Almeida, C. Wainwright, P. Mishkin, C. Zhang, S. Agarwal, K. Slama, A. Ray et al., Training language models to follow instructions with human feedback. Adv. Neural Inf. Process. Syst. **35**, 27730–27744 (2022)
16. B. Perrigo, Openai used kenyan workers on less than $2 per hour: Exclusive (2023). https://time.com/6247678/openai-chatgpt-kenya-workers/. Accessed 10 Feb 2023
17. C.R. Rogers, L. Carmichael, *Counseling and Psychotherapy: Newer Concepts in Practice* (Houghton Mifflin, 1942)
18. K. Shuster, M. Komeili, L. Adolphs, S. Roller, A. Szlam, J. Weston, Language models that seek for knowledge: Modular search & generation for dialogue and prompt completion (2022). arXiv preprint arXiv:2203.13224
19. K. Shuster, J. Xu, M. Komeili, D. Ju, E. Michael Smith, S. Roller, M. Ung, M. Chen, K. Arora, J. Lane, et al., Blenderbot 3: a deployed conversational agent that continually learns to responsibly engage (2022). *arXiv preprint* arXiv:2208.03188
20. N. Stiennon, L. Ouyang, J. Wu, D. Ziegler, R. Lowe, C. Voss, A. Radford, D. Amodei, P.F. Christiano, Learning to summarize with human feedback. Adv. Neural Inf. Process. Syst. **33**, 3008–3021 (2020)
21. R. Taori, I. Gulrajani, T. Zhang, Y. Dubois, X. Li, C. Guestrin, P. Liang, T.B. Hashimoto, Stanford alpaca: An instruction-following llama model (2023). https://github.com/tatsu-lab/stanford_alpaca
22. H. Touvron, L. Martin, K. Stone, P. Albert, A. Almahairi, Y. Babaei, N. Bashlykov, S. Batra, P. Bhargava, S. Bhosale, et al., Llama 2: Open foundation and fine-tuned chat models (2023). arXiv preprint arXiv:2307.09288
23. A. Turing, Computing machinery and intelligence. Mind **59**(236), 333–360 (1950)
24. M. Ung, J. Xu, Y.-L. Boureau, Saferdialogues: Taking feedback gracefully after conversational safety failures (2021). arXiv preprint arXiv:2110.07518
25. A. Vaswani, N. Shazeer, N. Parmar, J. Uszkoreit, L. Jones, A.N. Gomez, Ł. Kaiser, I. Polosukhin, Attention is all you need. Adv. Neural Inf. Process. Syst. **30** (2017)
26. R.S. Wallace, *The Anatomy of ALICE* (Springer, 2009)
27. J. Weizenbaum, Eliza-a computer program for the study of natural language communication between man and machine. Commun. ACM **9**(1), 36–45 (1966)
28. J. Xu, M. Ung, M. Komeili, K. Arora, Y-Lan Boureau, and Jason Weston. Learning new skills after deployment: Improving open-domain internet-driven dialogue with human feedback (2022). arXiv preprint arXiv:2208.03270

29. T. Zemčík, Failure of chatbot tay was evil, ugliness and uselessness in its nature or do we judge it through cognitive shortcuts and biases? AI & SOCIETY **36**, 361–367 (2021)
30. Y. Zhang, S. Sun, M. Galley, Y.-C. Chen, C. Brockett, X. Gao, J. Gao, JJ (Jingjing) Liu, B. Dolan, Dialogpt: Large-scale generative pre-training for conversational response generation (2019). arXiv:1911.00536
31. D.M. Ziegler, N. Stiennon, J. Wu, T.B. Brown, A. Radford, D. Amodei, P. Christiano, G. Irving, Fine-tuning language models from human preferences (2019). *arXiv preprint* arXiv:1909.08593

Chapter 3
Spontaneous Human Dialogues

3.1 Spontaneous Versus Prescriptive

It is a remarkable characteristic of human cognition that despite the seemingly haphazard nature of spontaneous conversations, we rarely perceive them as disordered or chaotic. Instead, our minds instinctively process, organize, and structure the information in real time, allowing us to build coherent mental models of the conversation. We effortlessly filter out redundancies, fill in gaps, and decode non-verbal cues, all while maintaining an active participation in the dialogue. This unconscious structuring is facilitated by our innate understanding of conversational norms, social context, and linguistic cues. Essentially, our brains constantly translate the raw, unpredictable flow of spontaneous speech into a meaningful and comprehensible narrative, allowing us to navigate the complexities of human communication.

Due to this automatic mental filtering, most people do not appreciate the complexity of the task and the incongruity of spontaneous speech. It is patently different from prescriptive written language, but it is also very different from scripted dialogues, both in audio and text layers. Spontaneous speech, as the term implies, is typically improvised and unplanned, full of idiosyncrasies, and often reflects the speaker's emotions, thoughts, and cultural background. It is immediate and ephemeral, characterized by interruptions, hesitations, repetitions, and non-standard grammar. On the other hand, prescriptive dialogue (e.g., scripted conversations) is usually planned, rehearsed, and follows established grammatical rules and conventions. It allows for reflection and revision, resulting in a more polished and formal mode of communication.

To understand the level of complexity of the task at hand, let us consider a real-world example of an unrefined raw ASR transcript of a spontaneous conversation between a customer and an agent presented in Fig. 3.1.

Several interesting properties of spontaneous speech can be seen in this transcript. Firstly, the overlap between the speaking parties is significant, in particular in the second part of the conversation, when the customer begins to exhibit anger. The raw

M. Morzy, *Spoken Language Processing*, Studies in Computational Intelligence 1205,
https://doi.org/10.1007/978-3-031-88566-2_3

CUSTOMER: yeah i have the appointment at ten this morning to nobody showed up
AGENT: okay apologize for the inconvenience one moment excited i'm showing here yeah they do you have this an early appointment uh one moment
CUSTOMER: okay
AGENT: just one moment well i looked at the condition or right at your door at ten according to what i have here and
AGENT: there was no answer the door
CUSTOMER: yeah i yeah i told to service technician that i spoke to last week when i studied it i gave him my garage code and gave him the instructions to send that technician into the garage and open it my garage
AGENT: yeah
CUSTOMER: code to be on the news
AGENT: oh yeah i understand
CUSTOMER: and it
AGENT: that are um it's
CUSTOMER: okay
AGENT: but unfortunately uh that ignition
CUSTOMER: no you
AGENT: are
CUSTOMER: only you only do it
AGENT: okay
CUSTOMER: that's
AGENT: that
CUSTOMER: your
AGENT: let me
CUSTOMER: job
AGENT: explain
CUSTOMER: you
AGENT: you
CUSTOMER: would do it
AGENT: the nations are not allowed to going to a property without the super uh
CUSTOMER: so
AGENT: owner

Fig. 3.1 Raw transcript of a spontaneous human-to-human conversation recognized by an ASR

text is really difficult to read and follow. Some intents appearing in the conversation are relatively easy to spot. The customer defines the main topic of conversation (*Appointment: I have the appointment at ten*) and states the problem (*Appointment Problem: nobody showed up*). The agent responds kindly (*Apology: apologize for inconvenience*) and stalls the conversation (*Buying Time: just one moment*). However, consider the intent *Anger*, which in this particular conversation is instantiated by the customer's utterance *"that's your job"*. Not only the instance of the intent is broken into three segments, but it is interwoven with examples of another intent when the agent tries to provide an explanation (*Explanation: let me explain*). In addition, the transcript is riddled with ASR errors. For instance, *"excited i'm showing"* is an erroneous rendering of *"because i'm shown"*. Other examples of mistakes include *"the condition or right"* which should be *"the technician were"*, *"studied"* being the wrong transcription of *"scheduled"*, *"that ignition"* (nonsensical in the context of the conversation) should read *"the technician"*, just as *"the nations"* should read *"the technicians"*. A very subtle ASR error is *"going to"*, where in fact the audio of the conversation contained the homophone *"go into"*. The example presented in Fig. 3.1 is a short snippet of a longer conversation, but clearly illustrates the challenges in analyzing spontaneous human conversations.

Hopefully, the reader needs no further convincing that spontaneous human speech is very different from prescriptive written language. But how different is it from prescriptive conversational language for which there exist large training corpora? Prescriptive conversational language can be found in movie dialogues, interviews, news broadcasts, recordings of theater performances, conversational parts of audiobooks, and televised debates. The common feature of these modalities is the fact that the dialogue is pre-scheduled, mostly devoid of the artifacts of spontaneous speech, and the degree of bi-directional conversation is often very limited. Spontaneous conversations often involve non-verbal cues such as body language, facial expressions, and tone of voice that add layers of meaning but are completely lacking from audio recordings. Verbal expressions of emotions are retained in the audio layer, but scripted dialogues usually use toned-down emotions, unlike in spontaneous speech. As Fig. 3.1 shows, interruptions and overlaps are common in spontaneous speech, whereas in scripted dialogues, each speaker typically completes their turn before the next one begins. Spontaneous conversations often include colloquialisms, slang, and non-standard grammar, features that tend to be edited or filtered from scripted dialogues that tend to adhere to proper grammar and language rules. Two features in particular make spontaneous speech difficult to parse and analyze: the presence of fillers and pauses, as well as corrections and clarifications introducing repetitiveness to the dialogue. Speakers in spontaneous conversations often repair their own speech errors or miscommunication. This may involve repeating themselves, correcting their grammar, or clarifying what they mean. Repairs are less common in scripted dialogues, where the speaker has had more time to plan their speech. Finally, spontaneous conversations are highly dependent on the context, including the relationship between the speakers, the situation, and the environment. Scripted dialogues are usually less influenced by these factors. In addition, spontaneous human conversations are also characterized by a sense of naturalness and authenticity. The speakers are typically engaged in a genuine exchange of ideas and information, and they are not trying to create a specific impression. Scripted dialogues, on the other hand, may be more artificial and performative, as the speakers are trying to stick to a pre-written script.

3.2 Artifacts of Spontaneous Speech

Let us now look closer into the artifacts of spontaneous speech that make this type of data unique and challenging. Artifacts can be categorized into audio artifacts, speech artifacts, and dialogue artifacts. Audio artifacts are elements that can be heard in the course of a conversation but are not part of the spoken words. These may include background noise, pauses, or variations in volume and tone that can affect the perception and understanding of the conversation. Speech artifacts are specific to the speaker's linguistic patterns, including the use of fillers, stutters, repetitions, or language-specific sounds. These can provide insights into the speaker's thought process and emotional state. Lastly, dialogue artifacts relate to the interaction between speakers. They encompass conversational elements such as interruptions, overlaps in speech, and turn-taking patterns, which can reveal dynamic aspects of the conversation like power structures, agreement, and rapport between the speakers.

3.2.1 Phonology and Phonetics

Before delving into audio artifacts of spoken conversations, we must clarify two concepts that are closely related to the processing of the audio layer of a conversation. The first concept is *phonology*, a subfield of linguistics that examines how languages arrange their sounds, or signs in the case of sign languages. Phonology is "the systematic use of sound to encode meaning in any spoken human language, or the field of linguistics studying this use" [5]. This term is also employed to investigate the sound organization of language variants. It analyses syllables, onset, rime, and mora at the level below a word. As it illustrates how sounds function to express meaning in a language, phonology is seen as one of the fundamental structures of any language. Some scholars categorize phonology under theoretical linguistics.

One of the important tasks of phonology is the study of *phonemes*, units comprising sounds that can be grouped into distinctive units. A true *phonetic transcription* represents a sequence of sounds and other speech qualities enclosed with square brackets. For instance, the proper pronunciation of the word *tie* could be transcribed as [t^{h}aɪ] *tie*, where the [^{h}] diacritic means that after the consonant [t], there is a release of air. These transcriptions are used only in specific cases when arbitrary precision is needed. In practice, a more useful tool to learn the pronunciation is the *phonemic transcription*, i.e., the sequence of phonemes. A *phoneme* is the smallest unit of sound in a language that can distinguish one word from another. It is an abstract concept that represents the range of sounds that are perceived as the same by speakers of a particular language, despite minor variations in pronunciation. The International Phonetic Alphabet (IPA) is the set of all characters used to graphically encode the sounds (*phones*) produced by human beings. IPA characters are used to create both phonetic and phonemic transcriptions [7]. Phonemes depend on the language they describe. Each phoneme is a set of sounds called allophones which, used interchangeably, do not change the meaning of the word. An *allophone* is a group of sounds that represent a single phoneme, or sound unit, in a specific language. The collection of sounds that make up an allophone varies from language to language, meaning a sound that is an allophone in Spanish might not be one in English. Native speakers typically do not consciously recognize allophones. This can create challenges when transcribing speech into the IPA, as different speakers might produce different sounds for the same phoneme. Although swapping a sound with an allophone generally does not alter a word's meaning, it can make the speaker sound non-native, or in extreme cases, render the word incomprehensible. One phoneme in two different languages, represented with the same IPA character, may describe different allophones (e.g., /a/ in the Arabic language has three allophones [ɑ], [ɐ] and [a], but in Russian /a/ may be mapped to [æ] or [ɑ]) [10].

The second subfield of linguistics relevant to our journey into spoken language processing is *phonetics*. Phonetics studies the physical sounds of human speech [11]. It is concerned with the physical properties of speech sounds or phonemes, their physiological production, acoustic properties, auditory perception, and neurophysiological status.[1] Phonetics has three main branches: articulatory phonetics, which examines the ways the vocal organs are used to produce speech sounds; acoustic phonetics, which focuses on the physical properties of speech sounds including pitch and volume; and auditory phonetics, which investigates how speech sounds are perceived by the human ear. The main task of phonetics is to provide a systematic and detailed description of the speech sounds in any given language and to understand how sounds function in particular languages and languages in general. Phonetics tries to understand two fundamental processes: the production of speech and the perception of speech. The main difference between phonetics and phonology is that they deal with speech sounds from different perspectives. Phonetics focuses on the physical aspects of sounds themselves while phonology examines the way sounds function within a particular language.

3.2.2 *Audio Artifacts*

Accent

The first audio artifact of speech is the accent. The term "accent" refers here to sociolinguistics, where it describes the way of pronunciation particular to a speaker or group of speakers, as opposed to phonetics, where the same term refers to the prominence given to a syllable in a word, or to a word in a sentence. It is a unique speech pattern that is influenced by various factors such as geographical location (a regional accent), social class (a social accent), ethnicity (an ethnolect), or first language (a foreign accent). A particular accent can influence various audio features within a person's speech. Most often it affects the pronunciation of consonants and vowels, which means that the same letter or combination of letters can sound distinctly different depending on the speaker's accent. An accent can influence the stress patterns within words and sentences, altering the rhythm and flow of speech. An accent can also change the intonation or pitch contour of speech, resulting in

[1] While phonetics traditionally deals with the physical properties of speech sounds, it is important to note that the principles of phonetics can also apply to sign languages. In this context, instead of vocal articulation and acoustics, the focus shifts to the specific physical movements and shapes made by the hands, body, and face. Thus, phonetics is not confined to audible languages but extends to visual-gestural languages as well.

unique melodic patterns that can carry different emotional or contextual meanings. Finally, accents can also involve specific phonetic features, such as the presence or absence of certain sounds, that are characteristic of a certain language or region. Usually, the grammar and vocabulary of a language do not change across different accents of a language, as the differences between accents lie predominantly in pronunciation. Broader linguistic differences lead to the concept of a dialect.

Dialect

A *dialect*, a term rooted in Greek words *diá*, meaning "through", and *légō*, meaning "I speak", refers to a specific form of a language that is unique to a certain group of speakers. Dialects create a spectrum and are not distinct from other variations of a language. Typically, a dialect is tied to a specific area (known as a regiolect), social class (referred to as a sociodialect), or ethnic group (termed as an ethnodialect). A dialect can be considered as standard or non-standard (vernacular), depending on several factors, including social, political, and cultural influences, historical context, prominence, and prevalence. More often than not, the term "dialect" is used to denote vernacular forms of a given language. If the linguistic difference is most pronounced in the lexical variations around vocabulary related to a particular profession or group, linguists define a given variety of a language as a *jargon* rather than dialect. *Cryptolects* are distinct vocabularies intentionally developed to keep outsiders at bay or to function as identifiers, examples of cryptolects include slang and argots. Finally, the unique speech patterns employed by a person are known as their *idiolect*.

Idiolects are one of many realizations of idiosyncrasies. The *idiosyncrasy* of speech refers to the unique characteristics or patterns of an individual's speech, which distinguishes them from others. These unique traits can include specific vocabulary usage, grammar, pronunciation, rhythm, speed, and even non-verbal cues such as gestures and expressions. These individual speech patterns are formed and influenced by various factors, including geographical location, social background, education, and personal experiences. The idiosyncrasy of speech helps in identifying a speaker and understanding their communication style but, at the same time, it significantly impedes the ability of statistical models to understand and transcribe human speech.

Noise

Finally, we must mention one more type of audio artifact that hinders the processing of audio recordings of conversations. Recall that spontaneous speech is unplanned, unrehearsed, and unscripted. The place of the recording is also often haphazard, the conversations are not recorded in a studio specifically designed to attenuate noise and emphasize speech, but in random locations. Background noise, such as ambient sounds, conversations, or traffic noises, can obscure the clarity of the speech being recorded. Acoustic echo, a common issue in recordings, is caused by sound reflecting off walls and other surfaces, creating a repeated or distorted sound. Similarly, electrical noise, resulting from poor-quality microphones or faulty wiring, can introduce a persistent hum or buzz. In many practical applications, in particular in scenarios involving call centers and customer support, audio is recorded over telephone landlines with low frequency. Wind noise, typically experienced in outdoor recordings, can create a heavy, distorted sound that muffles speech. These audio artifacts can make the speaker's words unclear, leading to inaccuracies and gaps in transcription. Advanced noise reduction technologies are often used to minimize the impact of these artifacts and to improve transcription accuracy.

3.2.3 *Speech Artifacts*

The second class of artifacts of spontaneous conversations are speech artifacts. These generally include phenomena related to the characteristics of unplanned speech. Here we will distinguish between general speech artifacts, and artifacts directly related to the structure of human-to-human conversations.

Sentence Segmentation

The first such artifact is the lack of proper segmentation which poses a significant challenge, particularly for speech processing and transcription tasks. Unlike written language, which has clear demarcations in the form of punctuation and spaces between words, spontaneous speech often flows continuously without explicit boundaries between words and sentences. Speakers may concatenate words, alter pronunciation in the flow of conversation, or use non-standard grammar and syntax, making it difficult to determine where one word or idea ends and another begins. Sentence segmentation can be encoded in prosody, where changing patterns of stress and intonation may indicate approximate boundaries of segments. This issue is compounded by factors such as speech speed, accent variations, and the presence of fillers and disfluencies. The absence of clear segmentation can lead to misunderstandings and inaccuracies in transcription, and it presents a complex problem for automated speech recognition systems, which must infer these boundaries to accurately convert speech into text.

Lack of Grammatical Structure

The second artifact of spontaneous speech is the lack of proper grammatical structure of spoken utterances. Spontaneous speech, particularly in informal and unscripted conversations, often lacks the rigid grammatical structure that is typically found in written language. Spontaneous speech often deviates from standard grammatical norms and is characterized by a certain level of informality and flexibility in grammar usage. For example, speakers may use sentence fragments, where a complete sentence is not formed. Another common occurrence is the use of run-on sentences, where multiple independent clauses are joined without appropriate punctuation or conjunctions, such as "I went to the store I bought milk." Speakers may also use incorrect verb tenses, as in "I seen that movie" instead of "I saw that movie," or they might use double negatives, which are not typically accepted in standard English, like "I don't need no help." Verb tenses and word order might not strictly adhere to grammatical rules. For example, a speaker might say, "You going store?" instead of the grammatically correct, "Are you going to the store?" These grammatical irregularities, while common in natural spoken language, can complicate the process of transcribing and understanding spontaneous speech.

3.2.4 Dialogue Artifacts

In the context of spoken language processing, the most interesting class of speech artifacts are the artifacts directly related to the conversational nature of human-to-human speech. These include a broad range of phenomena resulting from dialogue state control, expression planning, and turn-taking. An important class of dialogue artifacts is *speech disfluency*, the term that refers to any disruption, anomaly, or non-word utterance that interferes with the smooth progression of speech. Common forms include repetitions (repeating phrases or syllables), "false starts" (sentences that are abruptly interrupted), fillers, and self-corrections. Apart from their function in facilitating dialogue, speech disfluencies can also be used to analyze other elements of the conversation.

Fillers

The most noticeable form of speech disfluency are *fillers*, also known as *filled pauses* Fillers can constitute up to 20% of spontaneous speech [15]. They can serve as a means to pause for thought and plan the next part of the speech, in which case they are referred to as *planners* or *hesitation markers*. Hesitation markers in speech often occur due to the process of verbal planning or monitoring of the verbal signal. These can manifest as pauses, repetitions, false starts, and hesitation markers themselves [8]. Hesitation markers reveal that unscripted speech is impromptu. The conversation evolves organically as the speakers navigate through it, engaging in on-the-spot planning. A hesitation is a natural event that happens when a speaker is unsure of what to say next or how to articulate the forthcoming thought. Hesitation can take on various forms: filled pauses like "uhm", "umm", repetitions, corrections, and drawls. There is also a distinct group of "small words" that do not carry significant semantic weight in the conversation but serve a specific purpose in the speech flow. These words allow the speaker to gain time, such as "well, I mean", "like", "you know", "okay", or "basically", or serve as placeholders for a word that is momentarily elusive, like "stuff".

Filler words do not carry lexical content, they are strictly pragmatic conversation control mechanisms to mark the time to think without the intention of finishing one's turn of dialogue. The use of hesitation markers is necessary for maintaining the flow of conversation. They allow the speaker to indicate to the other party that she needs time to reflect without yielding control of the conversation to the other party. The capability to stall and signal the requirement for additional time to articulate thoughts using the right choice of words is especially important for individuals who are not native speakers of a language. Non-native speakers often require more time to structure and convey their thoughts due to more significant challenges in planning their speech.

It is noteworthy that speakers often reduce the fundamental frequency when uttering hesitation markers compared to the rest of the conversation. This behavior has been documented across multiple languages. The final problem with hesitation markers is the fact that they frequently go unrecognized and overlooked during a conversation, especially in native speech. This can result in these markers being absent from transcriptions, as they are subconsciously skipped by annotators. This is applicable to both parties in a conversation: the speaker is often unaware of the extent to which they use hesitation markers, and the listener tends to ignore these hesitations when interpreting the conversation's flow.

Hedging

A phenomenon closely related to hesitation markers is hedging. A *hedge* is a term integrated into a statement to denote hesitation, doubt, caution, likelihood, or vagueness. Hedges can enhance the courtesy of speech by reducing harshness and supporting politeness norms. They are typically adjectives or adverbs. In some cases, euphemisms can function as hedges. In written text hedges can be easily spotted, as in the following examples: "We have to consider just a few insubstantial problems" (the use of an adjective to diminish the seriousness of the discussed problems), "The weather was somewhat disturbed by an unexpected rain" (the use of an adjective to weaken the negative verb), or "He is responsible for this mess, isn't he?" (the use of the question clause to indicate hesitation). In verbal communication, hedges can be challenging as they act as markers of uncertainty, and their implied meaning often differs from their literal interpretation. Certain phrases in English serve distinct conversational functions to shape how statements are perceived or directed. For instance, phrases like "for all I know" and "to the best of my knowledge" are used to limit the force of a statement, indicating uncertainly or incomplete information. Similarly, the expressions "as I was told" and "according to her" limit the truthfulness of a statement by attributing the information to a third party, thus distancing the speaker from its veracity. To signal a change of topic, one might use phrases such as "by the way" or "actually," smoothly redirecting the conversation. Finally, to exhibit evasiveness, expressions like "as far as I can tell" and "I mean" are employed, implying indirect answers or hesitations.

Repetition

Another dialogue artifact is repetition, where a speaker repeats a word or phrase, usually unintentionally. Repetitions play a significant role in spontaneous speech, serving various communicative functions. While they are often regarded as errors or speech disfluencies, they can also be purposeful and beneficial. For instance, repetitions can be used by speakers to retain their turn in a conversation while they gather their thoughts or formulate more complex language structures. They can also be used for emphasis, to reinforce a point, or to ensure that the listener has understood the information correctly. Repetitions can be a part of a speaker's individual speech style or characteristic of certain dialects or cultural communication norms. On the listener's side, repetitions can aid comprehension and retention of information by providing additional processing time or reinforcing key points.

Overlap

Overlapping speech, where two or more speakers talk simultaneously, is also common in spontaneous conversations, making it difficult to accurately transcribe each speaker's words. Interruptions, false starts, and corrections are other artifacts that often occur. Non-verbal sounds such as laughter, sighs, or other paralinguistic features, while conveying meaning and emotion in face-to-face conversation, can be challenging to transcribe or interpret accurately in written form. A typical effect of overlapping speech is the presence of *back channeling*. Back-channeling refers to the phenomenon during a conversation where one participant interjects responses while the other is speaking. These responses serve a social function, such as indicating the listener's attentiveness, comprehension, or concurrence, rather than transmitting substantial information. Typical English back-channeling expressions are "OK", "right", "I see", and "yeah".

Phatic Expressions

Dialogues are ultimately a social experience. An important class of artifacts that appear in dialogues, but are not related to either surface or literal meaning of utterances, are *phatic expressions*. A phatic expression is a component of conversation that is used to uphold social interactions and relationships during a dialogue. Essentially, phatic expressions perform socio-pragmatic roles. They encompass social courtesies, expressions of respect, and displays of politeness. In certain contexts, the term "phatic expression" may also relate to the so-called small talk, a piece of conversation for its own sake. Also, the term "phatic expression" has been used to refer to communication functions within the dialogue, with notable examples including "I can't hear you", "you are muted", "you are breaking" during video calls. In general, phatic expressions play an important linguistic role in forming social connections between parties in a conversation. They create a sense of unity, closeness, and common understanding among participants, enhancing the overall pleasantness of the participants. The rise of digital communication gave birth to a new class of phatic expressions in the form of emojis, likes, re-tweets, and memes. Their role is also the expression of social bonding, co-participation, and social connection. Finally, there is a class of phatic expressions that cannot be captured by either audio or text representation of the conversation. These include non-verbal expressions such as posture, gesturing, waving, or direction of the gaze.

Discourse Markers

The last type of dialogue artifacts are *discourse markers*. Discourse markers are utterances that help to structure the flow of the dialogue [13]. They operate on the level of discourse, introducing relationships between individual utterances. Discourse markers usually do not change the truth-conditional meaning of sentences but serve to organize the flow of thought between utterances. There are several types of discourse markers, and each type plays a different role in the flow of conversation. Interpersonal markers serve as the means to organize and define the relationship between a speaker and a listener. They include utterances such as "look", "exactly", "you don't say!", "believe me". Referential markers mostly manage the causality between phrases using terms such as "then", "but", and "because". Structural markers build hierarchical structures between phrases, examples of structural markers include "firstly", "in the end", "finally", etc. The last type of marker bears a resemblance to hesitation markers and indicates the general flow of the speaker's thoughts. These markers often include phrases such as "I mean", "I just want to say that", "you know". Discourse markers are important in spontaneous speech as they help to structure the conversation, signal changes in topics, and facilitate smooth transitions between ideas. They can also indicate the speaker's attitude or emotional state.

3.3 Structure of Conversations

Despite being impromptu and unscripted, spontaneous human conversations have a certain degree of structure. Key areas of the analysis of this structure include dialogue acts, which shape the order and direction of conversations, topic-comment structure, which structures the assignment of meaning (what is being said and about what), speech acts, a key element in conveying intentions and achieving communication goals, linguistic expressions, the building blocks of the discourse, and turn-taking, a social mechanism that regulates the flow of dialogue. Each of these components plays a role in the understanding of how spontaneous conversations are built.

The analysis of language form, language meaning, and language in context is the focal point of linguistics. Notions discussed in this section are prevalent in both written text and spoken discourse. For instance, speech acts, which refer to the actions performed via utterances, like making a request or giving an order, are fundamental to all forms of communication. They highlight how language is used to perform actions and convey intentions, in both written and spoken modes. Similarly, linguistic expressions, which encompass words, phrases, or sentences that communicate specific meanings, are integral to the creation and interpretation of both written texts and spoken conversations. These tools collectively contribute to our understanding of how language works, irrespective of the medium. Here we focus on the spoken language, but these notions are equally important for traditional written text.

3.3.1 Dialogue Acts

A *dialogue act* is a concept in linguistics and computer science that refers to the meaning or purpose of a segment of speech within the context of a conversation. Rather than focusing on the literal meaning of words or sentences, dialogue acts consider the intentions or functions behind the utterances. For example, a statement could be an assertion, a question could be a request for information, and an utterance like "Could you pass the salt?" is a form of request. Identifying dialogue acts is important in natural language processing and understanding, as it helps in building systems like chatbots or voice assistants that can understand and respond to human conversation in a more context-aware and natural manner. The classification of spans of dialogue into dialogue acts is known as *spoken utterance classification*. Often, there is a fixed set of tags of dialogue acts, typical examples include "statement", "opinion", "yes-no-question", or "accept/agree". Dialogue act may represent a particular function of an utterance, but it may be also a part of the structure of the conversation (e.g. "salutation", "closing").

Classification of spans of spontaneous conversation into dialogue acts is a challenging task. First, context plays an important role in interpreting the nature of dialogue acts, and without sufficient understanding of the surrounding conversation, the intent behind an utterance may not be clear. Secondly, people often use indirect language, sarcasm, or cultural references, which can be difficult to correctly classify without a deep understanding of the language and culture. A single utterance can often serve multiple functions, making it hard to assign it to a single dialogue act category. The artifacts of spontaneous speech discussed previously, such as interruptions, overlaps, and non-verbal cues, make the task of spoken utterance classification even more complex.

Over the years, many different ontologies of dialogue acts have been proposed and defined. Examples of such ontologies include the Dynamic Interpretation Taxonomy DIT++ [1], Dialogue Act Markup using Several Layers DAMSL [6], and Dialogue

Act Markup Language DiAML [4]. The common feature of these ontologies is the use of several dimensions based on which particular tags of dialogue acts are defined. The most common set of dimensions is the following:

- ask/activity: the span performs a task or an activity in the dialogue
- auto-feedback: the span contains information about the processing of the previous speaker's utterance
- allo-feedback: the span contains the speaker's opinions or elicits information about the addressee's processing of previous utterances
- contact management: the span serves the purpose of maintaining contact in conversation
- turn management: the span is related to the turn-taking, i.e., it contains utterances that try to take over, keep, yield, or accept the role of the speaker
- time management: the speaker is buying time to better express themselves
- discourse structuring: the span is either the start or the closing of the conversation
- own communication management: the span performs a correction of the speaker's utterance
- social obligations management: the span contains markers of social conventions, such as greetings, apologies, and expressions of gratitude.

These dimensions have been introduced by Bunt [3] based on the following observation: "Besides trying to perform the underlying task or activity that motivates the dialogue, he [the user] has to monitor mutual attention and correctness of understanding; he furthermore has to manage a variety of aspects of the interaction, including the distribution of sender and receiver roles ('turn management'), the continuation of the interaction under time constraints ('time management'), and the introduction and closing of topics ('topic management'); and he also has to deal with social obligations such as introducing oneself, apologizing, thanking and greeting." Independently of this set of dimensions, there are general purpose functions:

- information seeking functions: yes/no questions, WH-questions (where? who? what? why? when? which? how?), multiple choice questions
- information providing functions: informs, agreements, disagreements, corrections, answers, confirmations
- commissive functions: committing to an action, such as offers, promises, acceptance of requests
- directive functions: speaker wants the addressee to consider an action, such as instructions, requests, suggestions

To further complicate things, dialogue act dimensions can be grouped into four main layers of annotation. Forward- and backward-looking layers represent communicative functions. The information status layer annotates utterances that are impossible to annotate (e.g. inaudible utterances). Information level annotations define utterance as being about a task, management of a task, or management of communication.

Dialogue utterances are multi-functional, they serve different purposes during the management of the dialogue. Thus, dialogue act annotation must take this into consideration and provide multiple tags to dialogue acts. The annotation schemes such as DAMSL and DiAML allow for multidimensional annotation, but the selection of tags is questionable, some tags are mutually exclusive while others imply other tags. Existing annotation schemes also do not make a clear distinction between notions of a dimension and a layer. As Bunt notes [3]: "Many communicative functions (questions, statements, answers, requests) do not belong to any dimension, but can be used in any dimension as general-purpose language functions, depending on the semantics of the utterance."

3.3.2 Topics, Comments, Speech Acts, and Linguistic Expressions

In linguistics topics and comments are used to describe the structure of a sentence or utterance. The *topic* (also known as *theme*) is the part of the sentence that the rest of the sentence, the *comment* (also known as *rheme* or *focus*), is about. The topic is often the subject of the sentence, but not always, and the comment provides new information about the topic. Both topic and comment are parts of the sentence but at the pragmatic rather than syntactical level. For instance, in the sentence "The large black dog is barking loudly", "the large black dog" is the topic, and "is barking loudly" is the comment. The comment gives more information or makes a statement about the topic. In another example, "The movie I saw yesterday was very boring", the part "The movie I saw yesterday" is the topic, and "was very boring" is the comment. Here, the comment is expressing an opinion on the topic. A topic is different from the subject of the sentence. A subject is defined by the syntax and grammatical rules. A topic is defined by the context which provides the meaning of the sentence.

The next level of sentence analysis involves the identification of speech acts. A *speech act* is the action performed by a speaker through the use of language [2]. It is based on the idea that when we communicate, we do not merely convey information but also perform actions, such as making requests, giving orders, or making promises [14]. Speech act theory emphasizes the functional and purposeful nature of human communication, asserting that a speech act can be understood in three ways: the locutionary act (the actual utterance and its apparent meaning), the illocutionary act (the intended meaning or action), and the perlocutionary act (the effect of the speech act on the listener).

Two important notions are closely related to the speech act theory, namely, the concepts of intention and proposition. *Intention* refers to the speaker's purpose or goal in making a particular utterance. It is closely tied to the illocutionary act. For example, the intention behind the statement "It's cold in here" might be to request that someone close the window. *Proposition*, on the other hand, refers to the content or meaning of the utterance. It is associated with the locutionary act. An example of a proposition is the statement "When viewed above the North Pole, the Earth rotates counterclockwise, from west to east." Here, the speaker is making an utterance that has a specific propositional content or meaning, i.e., stating factual information about the direction of the Earth's rotation. The act of saying this statement itself is the locutionary act, and the proposition or the factual content conveyed is that the Earth has a prograde rotation. Both intention and proposition have a profound impact on language understanding, it is the recognition that the meaning of a speech act is not solely determined by the literal meaning of the words (proposition) but also significantly by the speaker's intention. Understanding communication requires an understanding of both the context and the speaker's intentions.

The last level of the analysis of conversation structure is linguistic expression. A *linguistic expression*, in the broadest sense, refers to a sequence of words, sentences, or text that conveys a particular meaning or idea. It encompasses all components of language that can be used to express information, thoughts, or emotions. A linguistic expression can be a single word, such as "sad," which conveys a specific emotion. A linguistic expression might be a statement like "I am eating breakfast," which communicates a particular action being performed by a subject. At the text level, a linguistic expression could be an entire paragraph, where a series of sentences work together to convey more complex thoughts or narratives. One should note that linguistic expressions are not just about the literal meanings of words and sentences, but also about how context, tone, and other factors contribute to the conveyed meaning.

All notions mentioned in this paragraph are linguistic tools to convey meaning using language. Usually, topics, speech acts, or linguistic expressions are discussed in the context of written text. Identifying these phenomena in spontaneous speech is more challenging. Topics and comments change constantly throughout the conversation, and matching them in transcripts is non-trivial. Similarly, speech acts can change rapidly and without clear demarcation due to the spontaneous nature of conversation. The intention of the speaker, which is required to determine the type of speech act, may not always be explicitly stated in utterances and might be conveyed through subtleties or nuances in tone or body language, features that are lost in transcripts. Identifying linguistic expressions in human conversations can also be complex. It requires an understanding of idiomatic expressions, cultural references, and colloquial language, which is far more common in spontaneous dialogues. Also, speakers sometimes use incomplete sentences and change topics abruptly, which can make it hard to isolate and identify specific linguistic expressions.

3.3.3 Turn-Taking

The dialogue consists of overlapping turns of speaking parties, but the exchange of control over the conversation has to be managed and coordinated. *Turn-taking* is the organization of the conversation into a sequence of alternating turns. Each turn either introduces a new topic to the conversation or responds to the previous turn. The decision on the timing of transitions may depend on both linguistic and non-linguistic cues. Turn-taking is one of the aspects of spontaneous human conversations that are most influenced by the culture because it is the culture that defines the norms for the amount of speech overlap that is allowed (what is the threshold beyond which the overlap is considered rude and hostile take-over of the control of conversation), the expected length of pauses between turns, what signals of transition are acceptable, and what is the expected distribution of turns of participants given their age, gender, social status, etc.

Turn-taking is defined by three components. The first component is responsible for constructing turn units, basic blocks of conversation carrying its meaning. The second component defines the technique to select the next speaker (which is a fascinating problem in a multi-speaker conversation), and the third component defines the rules for building turn units and designating the next speaker. The main objective of turn-taking is to minimize the amount of silence within the conversation, at the same time ensuring the civility of the conversation. Unfortunately, there are no universal mechanisms to guarantee turn-taking fairness, such as equal distribution of turns between speaking parties.

The most difficult element of turn-taking is the management of overlap. One taxonomy identifies four different types of conversational overlap:

- terminal: a speaker assumes that the previous speaker is concluding her utterance and starts their own turn,
- back-channeling: a speaker uses verbal cues to signal attention without the willingness to take over the turn,
- conditional access: a speaker permits a short input from another speaker to enrich the conversation without officially handing over the speaking turn,
- chordal: both speakers speak at the same time without animosity, for instance, laughing simultaneously.

Alternatively, turn-talking can be characterized as transitional (a speaker starts an utterance at a possible end of the previous utterance), recognitional (a speaker anticipates the end of the previous utterance and tries to finish it for the current speaker), and progressional (a speaker uses a speech disfluency of the previous speaker to take over the turn). Each of those types of overlaps is heavily codified by cultural norms of polite conversation, but also by individual factors such as the speaker's temperament, attitude, current mood, degree of involvement in the conversation, and many more.

The problem of turn-taking has been the subject of an interesting modeling attempt. Most current models struggle with producing natural-sounding dialogues, and the turn-taking is notoriously hard to model. Conversation markers of turn transition consist mainly of either non-verbal cues, or verbal cues that are not proper words and thus are omitted from language model vocabulary. Also, many approaches try to model dialogue from text inputs, where these cues are not present. Nguyen et al. present a method to generate naturally sounding (although not semantically correct) turn-taking in conversations [12]. The method uses only audio inputs, without the need to have aligned transcripts. The audio output is produced by a new model: the Dialogue Transformer Language Model (DLG). Because training is performed directly on audio inputs, the model can learn much more useful features for mimicking natural turn-taking, and, in particular, it can generate naturally sounding overlaps. The resulting audio is quite similar to human face-to-face conversations. The authors train the model directly from audio, bypassing ASR and text. Speech units represent both verbal and non-verbal inputs. The novel dual-tower transformer architecture passes each channel through a separate tower, and the towers are joined by a cross-attention transformer layer which produces "hints" about what is happening in the opposite channel of conversation. The transformers work on wave2vec representations extracted from the audio. Each channel is tokenized via a separate HuBERT transformer [9], a self-supervised model that generates audio representations. Tokenization is performed by a convolutional neural network and its result is passed to a transformer. These representations are sent to HiFi-GAN to generate audio (separately for each channel). New tokens are predicted based on the input streams, and the new cross-attention layer joining both channels allows for generation overlaps that sound natural. The most important improvement is the loss function which consists of two components. The *edge unit prediction objective* forces the model to predict the next unit only if it is different from the current one (i.e. edge unit) via cross-entropy loss, and the *duration prediction objective* forces the model to learn the duration of edge units via L1 loss. The quality of the model has been evaluated by crowdsourcing and the naturalness of turn-taking has been rated on a 1–5 scale. Ground truth rating was 4.25 for both naturalness and meaningfulness. DLM resulted in 3.86 naturalness and 2.71 meaningfulness, outperforming a traditional Transformer-based model. DLM has been trained on the Fisher dataset with 16 000 English telephone conversations, 2000 hours of transcribed speech, and separate audio channels.

3.4 Summary

In this chapter, we have explored the characteristics of spontaneous human dialogues, highlighting their inherent complexities and the distinct differences from written text. Spontaneous conversations are marked by their impromptu nature, leading to numerous artifacts at the audio, speech, and dialogue levels. These include hesitations, interruptions, repetitions, and non-standard grammar, which are influenced by emotional states, cultural backgrounds, and contextual factors. Spontaneous dialogues often lack clear segmentation and grammatical structure, making them challenging to parse and understand. Overlapping speech and turn-taking add to the complexity, requiring advanced understanding and modeling of conversational dynamics. These features, alongside the implicit structure of dialogue acts and speaker turns, suggest the need for a new blend of artificial intelligence and linguistics to achieve true spoken language understanding.

References

1. Dit++ taxonomy of dialogue acts, annotation scheme, and diaml markup language. https://dit.uvt.nl. Accessed 26 Oct 2023
2. J.L. Austin, *How to do Things with Words*, vol. 88 (Oxford University Press, 1975)
3. H. Bunt, Dimensions in dialogue act annotation, in *LREC* (2006), pp. 919–924
4. H. Bunt, J. Alexandersson, J.-W. Choe, A.C. Fang, K. Hasida, V. Petukhova, A. Popescu-Belis, D.R. Traum, Iso 24617-2: A semantically-based standard for dialogue annotation, in *LREC* (2012), pp. 430–437
5. J. Clark, C. Yallop, J. Fletcher, *An Introduction to Phonetics and Phonology* (Blackwell Textbooks in Linguistics, Wiley, 2007)
6. M.G. Core, J. Allen, *Coding dialogs with the damsl annotation scheme, in AAAI Fall Symposium on Communicative Action in Humans and Machines*, vol. 56 (MA, Boston, 1997), pp.28–35
7. P.T. Daniels, W. Bright, *The World's Writing Systems* (Oxford University Press, 1996)
8. G. Gilquin et al., Hesitation markers among efl learners: Pragmatic deficiency or difference. Pragmat. Corpus Ling. Mutual. Entente **2**, 119–149 (2008)
9. W.-N. Hsu, Y.-H.H. Tsai, B. Bolte, R. Salakhutdinov, A. Mohamed, Hubert: How much can a bad teacher benefit asr pre-training? in *ICASSP 2021-2021 IEEE International Conference on Acoustics, Speech and Signal Processing (ICASSP)* (IEEE, 2021), pp. 6533–6537
10. D. Jones, D. Ward, *The Phonetics of Russian* (Cambridge University Press, 2011)
11. J. Laver, *Principles of Phonetics* (Cambridge University Press, 1994)
12. T.A. Nguyen, E. Kharitonov, J. Copet, Y. Adi, W.-N. Hsu, A. Elkahky, P. Tomasello, R. Algayres, B. Sagot, A. Mohamed et al., Generative spoken dialogue language modeling. Trans. Assoc. Comput. Linguist. **11**, 250–266 (2023)
13. D. Schiffrin, *Discourse Markers*, vol. 5 (Cambridge University Press, 1987)
14. J.R. Searle, *Speech Acts: An Essay in the Philosophy of Language*, vol. 626 (Cambridge University Press, 1969)
15. J.E.F. Tree, The effects of false starts and repetitions on the processing of subsequent words in spontaneous speech. J. Memory Lang. **34**(6), 709–738 (1995)

Part II
Transcripts

Automatic Speech Recognition (ASR) systems transcribe human speech into written text, and the resulting transcripts present their own unique complexities. The transcripts produced by ASRs can contain disfluencies like stutters, repetitions, and filler words, which are common in spontaneous speech but rare in written text. Additionally, the structure of these transcripts is typically less formal, with sentences frequently left unfinished or grammatically incorrect, and dialogues often sprinkled with colloquialisms, slang, and jargon.

The complexity of ASR-produced transcripts arises from the limitations of the technology. ASR systems, while sophisticated, can struggle with homophones, accents, speech impediments, and background noise, often leading to incorrect transcriptions. They also lack the ability to understand context, which can cause errors in transcribing words that sound similar but have different meanings depending on their usage. The output of an ASR system (tokens appearing in the transcript) is limited to the built-in vocabulary of the ASR, so many named entities, such as personal names, geographical locations, company names, and product names, are approximated by a similar sounding word from the dictionary, producing a word error as the result. Moreover, ASR systems do not capture non-verbal cues, like tone, pitch, or pauses that can significantly alter the meaning of an utterance. Despite these challenges, ASR technology continues to improve and plays an increasingly important role in transcribing human conversations for applications ranging from customer service to healthcare.

Chapter 4
Automatic Speech Recognition

4.1 What is Automatic Speech Recognition?

Automatic Speech Recognition serves as a bridge between human speech and computer understanding. It is a technology that converts spoken language into written text, using computational linguistics and machine learning. This technology lies at the intersection of Text-to-Speech (TTS) and natural language processing. While TTS translates written text into spoken language, allowing computers to communicate audibly with users, ASR does the reverse, enabling computers to comprehend and transcribe human speech. The transcriptions generated by ASR can then be processed and analyzed using NLP, giving computers the ability to understand the semantic and syntactic structure of the human language along with the intent, sentiment, and context, enhancing human-computer interaction.

The complexity of ASR stems from the hierarchical nature of the problem and the variability of the sources it deals with, namely acoustics, phonetics, and linguistics. Acoustically, speech is a continuous, dynamic signal with a wide range of frequency and amplitude variations. It is affected by speaker characteristics such as age, gender, emotional state, and accent, as well as environmental factors like background noise or microphone quality. On the phonetic level, the same phoneme can sound different depending on its surrounding phonemes (which is known as co-articulation), and different phonemes can sound similar, leading to possible confusion. Another phenomenon often appearing in spontaneous speech is elision, which is the omission of one or more sounds, syllables, or words in a spoken phrase. It is a common feature in natural speech and it occurs due to various factors, including ease of pronunciation ("don't" → "don"), speed of speech ("I am going to" → "I'm gonna"), and the influence of surrounding sounds ("oranges and apples" → "oranges an apples", "history" → "histry"). Elision can make the spoken language more natural and fluent, but it can also make it more difficult to understand, especially for non-native speakers. Linguistically, languages have complex and often ambiguous structures.

M. Morzy, *Spoken Language Processing*, Studies in Computational Intelligence 1205,
https://doi.org/10.1007/978-3-031-88566-2_4

Words can have multiple meanings depending on context, and phrases or sentences can be interpreted in different ways.

From the point of view of machine learning, ASR also faces additional challenges due to its high-dimensional output space and the complex sequence-to-sequence nature of the problem. It has to map sequences of acoustic signals to sequences of phonetic units, words, or sentences, which is a highly complex task due to the variability, non-linear nature of speech, and the huge size of the output vocabulary. Furthermore, there is often a lack of adequately annotated training data for ASR systems. While large amounts of speech data exist, it is time-consuming and expensive to annotate this data accurately with transcriptions, which are needed for supervised learning approaches in ASR. Lastly, the data used for ASR is often noisy. This noise can come from various sources like background sounds, overlapping speech, room reverberation, speaker accents, or errors in the transcription used for training. These factors, combined with the speed and spontaneity of natural speech, make ASR a highly challenging field.

In the context of ASR, several related but distinct terms often come into play. Speaker diarization refers to the task of determining 'who spoke when,' separating and attributing speech segments in an audio stream to different speakers. Speaker recognition, on the other hand, involves identifying or verifying a speaker's identity using characteristics extracted from their speech. Sentiment analysis from audio involves determining the emotional tone or sentiment conveyed in a spoken message. While these tasks are closely related to ASR, they are not synonymous with it. ASR focuses on transcribing spoken language into written text, while speaker diarization, speaker recognition, and sentiment analysis provide further analysis of the speech. Collectively, these tasks, combined with others like speech synthesis and natural language processing, contribute to what is sometimes referred to as Spoken Language Understanding (SLU).

4.2 Brief History of ASR

Large part of the history of ASR systems has been already covered in Sect. 2.4, so here we will briefly summarize the long path that led from systems capable of recognizing a single spoken digit to contemporary systems capable of transcribing speech in 100 different languages. The first epoch of ASR systems traces back to the 1950s with the pioneering work of Bell Labs. They developed a system which could recognize spoken digits from 0-9, illustrating the initial feasibility of speech recognition technology [27]. Around the same period, IBM introduced the Shoebox system, which recognized 16 English words and was capable of performing arithmetic operations based on voice commands.

Two significant scientific developments during this period further shaped the evolution of ASR systems. The first was the publication of the Source-Filter Model of Speech Production by Fant [15], which provided a theoretical foundation for understanding and simulating the process of human speech production. This model has influenced the feature extraction methods used in ASR. The second development was the introduction of the Dynamic Time Warping (DTW) technique [53]. DTW technique is a powerful algorithm used in time series analysis, particularly useful in the context of ASR. The fundamental principle of DTW is to align two temporal sequences that may vary in speed and length, making it optimal for dealing with the temporal variability present in spoken language. DTW calculates an optimal match between two sequences of feature vectors (e.g., speech signals) by warping the time dimension in a non-linear fashion. It achieves this through a process that minimizes the distance between the two sequences, allowing for stretching or compressing of the time axis. This ability to handle time-dependent variations makes DTW an important tool in ASR, enabling the system to accurately align and compare spoken words or phrases despite differences in speech rate, duration, or elongation.

Between 1970 and 1990, the field of ASR saw the development of several influential systems. The main driver behind the acceleration of the research was Speech Understanding Research (SUR), a 5-year program funded by the Defense Advanced Research Projects Agency (DARPA). Initiated in the 1970s, this program aimed to develop systems capable of understanding continuous speech from a wide vocabulary. Several universities and research institutions participated in the SUR program, leading to the creation of some of the first large-vocabulary ASR systems. The funding and direction provided by DARPA through the SUR program significantly accelerated the progress in ASR and laid the foundation for many of the speech recognition technologies in use today. The DARPA-sponsored annual evaluations also established a benchmark for progress in the field. Among the direct results of the SUR program were systems such as Tangora, Dragon Dictate, and Sphinx.

IBM's Tangora, launched in the early 1980s, was the first large-vocabulary, speaker-dependent ASR system, capable of recognizing about 20 000 English words. This marked a considerable leap from the previous limited-vocabulary systems. Later in the 1980s, the Apricot Portable, developed by Dragon Systems, became one of the first attempts to incorporate ASR technology into a portable device. In 1990, Dragon Systems made another major contribution with the release of Dragon Dictate, a groundbreaking system that introduced continuous speech recognition, allowing users to speak naturally instead of pausing between words. While it was initially slow and required a powerful computer, Dragon Dictate paved the way for future developments in continuous, large-vocabulary ASR. Another system developed during this period was the Sphinx by Carnegie Mellon University, one of the first open-source ASR systems. The availability of Sphinx as open-source software led to rapid advancements in ASR by facilitating research and allowing many developers to contribute to improving speech recognition technology.

At this time, the main scientific development was the application of Hidden Markov Models (HMMs) to speech recognition [26]. Pioneered by Janet and James Bakers in the late 1970s and early 1980s, HMMs represented a significant milestone in the evolution of speech recognition. HMMs are particularly well-suited for speech recognition because they effectively handle time sequence data, such as speech, where the current state depends on previous states. In an HMM, the system being modeled is assumed to be a Markov process with hidden states, meaning the states themselves are not directly observable, but the output dependent on the state is observable. For ASR, this allows the system to model the probability of a sequence of observed acoustic vectors (speech signal) given a sequence of hidden states (words or phonemes). The introduction of HMMs led to a dramatic improvement in speech recognition performance and became the dominant approach in ASR systems for several decades. It enabled the creation of more accurate and robust speech recognition systems capable of dealing with real-world variability in speech.

The 2000s saw dramatic advances in ASR, with several new systems emerging. DARPA continued to fund research into speech recognition launching two speech recognition programs: Effective Affordable Reusable Speech-to-Text (EARS) in 2002 and Global Autonomous Language Exploitation (GALE) in 2005. The 2000s marked the beginning of the large-scale utilization of speech recognition by intelligence agencies funding several programs to advance the potential of ASR for espionage and counterintelligence applications. One of those initiatives was the Babel Program, funded by the Intelligence Advanced Research Projects Activity (IARPA). Babel aimed to develop methods for rapid language adaptation in ASR systems. The goal was to quickly build and deploy effective ASR systems for any human language, enabling intelligence officers to transcribe and analyze multilingual speech data efficiently. More importantly, ASR systems have reached the maturity that allowed their deployment to the consumer market. In 2001, Microsoft introduced speech recognition capabilities into Office XP, making ASR more accessible to everyday users. Google launched Google Voice Search (also referred to as GOOG-411) in 2008, a mobile app that used cloud-based ASR to transcribe spoken search queries. Another key development was IBM's Watson system, which used ASR technology to compete on the quiz show Jeopardy. Meanwhile, in 2011, Apple's Siri was released, bringing speech recognition technology into the mainstream. Siri, a voice-activated assistant, used ASR to enable users to interact with their devices using natural language. This marked the beginning of widespread consumer use of ASR technology.

Two important developments in ASR technology took place in the 2000s: the advent of Long Short-Term Memory (LSTM) networks and the introduction of Deep Neural Networks (DNNs) into acoustic modeling. LSTM networks, a type of Recurrent Neural Network (RNN) invented by Hochreiter and Schmidhuber [25] in 1997, were first applied to speech recognition in the early 2000s [38]. Unlike traditional feed-forward neural networks, LSTM networks have feedback connections and can process sequences of data, making them particularly suited for time-series data like speech. LSTMs are designed to overcome the vanishing gradient problem in traditional RNNs, allowing them to capture long-range dependencies in the input data. This makes them highly effective for tasks such as ASR, where the recognition of a current sound may depend on the context of sounds spoken much earlier.

The second significant development was the introduction of DNNs into acoustic modeling by Geoffrey Hinton and his students in the mid-2000s [24]. By using multiple layers of non-linear processing units, DNNs could model complex patterns in the data, improving the accuracy of speech recognition. The ability of DNNs to learn hierarchical representations—where each layer of the network learns increasingly complex features—proved very powerful in modeling the complexities of human speech. This was a paradigm shift in ASR, with DNNs quickly becoming a standard component of modern speech recognition systems.

The decade following 2010 was decisive in the history of ASR, with several advancements in models, techniques, and solutions. One of the most transformative developments was the widespread adoption of Convolutional Neural Networks (CNNs) for acoustic modeling. CNNs led to a visible improvement in ASR, outperforming traditional Gaussian Mixture Model-Hidden Markov Model (GMM-HMM) systems. Another important development has been the development of end-to-end ASR systems. End-to-end ASR systems directly map from the audio signal to the text transcript, without the need for intermediate representations such as acoustic and language models. This has simplified the ASR training process and made it possible to train ASR models on much larger datasets. In 2014, Baidu introduced Deep Speech [22], an end-to-end ASR system built entirely using deep learning. Meanwhile, technologies such as Google's Voice Search and Apple's Siri became increasingly sophisticated, driving the consumer adoption of ASR. Another key development was the application of attention mechanisms [58], initially introduced in the field of natural language processing, to ASR. This led to the adaptation of the Transformer model to ASR tasks. The rise of cloud computing and the proliferation of smart devices have facilitated the deployment of ASR technology on a much larger scale, allowing for real-time, on-device speech recognition.

4.3 Fundamentals of ASR

4.3.1 Speech Production

The human speech production apparatus is a complex system involving various organs and structures. The process begins with the lungs, which provide the airflow necessary for speech. The air is then passed through the glottis, the space between the vocal cords located in the larynx or the voice box. The vocal cords vibrate to modulate this airflow, creating the basic sound elements of speech. The shaping of this sound into recognizable speech sounds, or phonemes, occurs in the vocal tract, which includes the pharynx, oral cavity, and nasal cavity. The pharynx connects the larynx to the oral and nasal cavities and contributes to the resonance characteristics of speech sounds. The tongue, one of the most flexible parts of the speech apparatus, changes its position and shape to articulate different phonemes. The teeth, lips, and hard palate further refine the sounds by providing surfaces against which the tongue and soft tissues can create constrictions or closures, forming different speech sounds. Finally, the nose plays a part in speech production by providing an alternate path for the airflow, which is critical in producing nasal sounds like *m*, *n*, and *ng*. By understanding these physiological processes, researchers can design ASR models that better mimic human speech patterns and improve their accuracy in recognizing spoken words and phrases. This understanding can aid in the development of feature extraction methods, which identify distinctive characteristics in the speech signal that are useful for recognition. It can also inform the design of acoustic models, which map these features to speech units.

Although the physical process of speech production involves complex interactions between the above-mentioned components, it can be successfully simplified as a mathematical model. The most influential speech production model has been developed by Fant [15] and Stevens [55] under the moniker of the *source-filter model*. The source-filter model represents speech as the superposition of two independent components. The *source* is the generation of a raw sound signal, primarily produced by the vibration of the vocal cords in the larynx. The pitch and loudness of the sound are determined at this stage. The *filter* refers to the modifications to this raw sound as it travels through the vocal tract (including the pharynx, oral cavity, and nasal cavity). The specific shape and configuration of the vocal tract at any moment — controlled by the positions of the tongue, lips, jaw, and other articulators — determine the resonant frequencies of the vocal tract, which in turn selectively amplify or attenuate components of the source signal. The result is the complex mix of frequencies that we perceive as speech. Assuming that the filter is linear and time-independent, this convolution can be expressed as follows:

$$s(t) = \int_{-\infty}^{\infty} e(t) f(t - \tau)\, d\tau \tag{4.1}$$

where s is the speech, e is the source (excitation), and f is the filter. The integral sums the product of the excitation signal and a time-shifted version of the filter's impulse response, effectively combining them to produce the output signal. Equation 4.1 is fundamental in speech processing, where it models how sound waves are generated and modified by vocal tract dynamics.

Unfortunately, this model is a simplification. The true process of speech production is more elaborate. Just by looking at different frequency spectra, we can observe the following. The lungs expel air, creating an airflow that travels upwards towards the larynx. This airflow is the initial waveform that forms the basis of speech production. The pressure of the airflow causes the vocal cords to vibrate. This vibration of the vocal cords modulates the airflow, turning the continuous stream of air into a series of pulses. Each pulse of air corresponds to a cycle of the vocal cord vibration, and the rate of these pulses determines the pitch of the voice. The lowest frequency of this periodic waveform (the rate at which the vocal cords vibrate during phonation) is called the *fundamental frequency* in speech production and is denoted as $F0$. The faster the vocal cords vibrate, the higher the fundamental frequency and the higher the perceived pitch of the voice. Conversely, slower vibration results in a lower fundamental frequency and a lower voice pitch. It is worth noting that the fundamental frequency can vary greatly depending on factors such as the individual's sex, age, and emotional state. For instance, emotional states like excitement or fear can cause an increase in the fundamental frequency, leading to a higher-pitched voice.

The pattern of pulses produced by vocal cords creates what is known as the *glottal spectrum*, the raw sound produced by the vocal cords. The glottal spectrum then travels through the vocal tract, which acts as a filter, shaping the glottal spectrum by selectively amplifying or dampening certain frequencies based on its current shape and configuration. The specific resonances of the vocal tract are called *formants*, and these formants emphasize certain frequencies over others. For example, the positions of the tongue, lips, and jaw can change the shape of the vocal tract and therefore the positions of the formants, altering the sound that is ultimately produced. The final output of this process is the spectrum of sound frequencies that we perceive as speech.

4.3.2 Statistical Language Modeling

ASR systems could, in theory, be trained to recognize each word in a large vocabulary. However, this approach would not be reasonable. The amount of training data required would be enormous. Each word would need to be represented in the training set in a variety of accents, dialects, and contexts, which would make the data collection process unfeasible in practice. Also, language is highly dynamic, with new words and phrases constantly being introduced. A model trained to recognize a fixed set of words would struggle to adapt to linguistic changes over time. Finally, the task of differentiating between 50 000 words (typical size of a large language model vocabulary) with hundreds of homophones (words that have the same pronunciation but different meaning or spelling, such as "for" and "four", or "sea" and "see") would be extremely difficult.

This is why all modern ASR systems are trained to recognize phonemes, which we discussed in Sect. 3.2.1. Typically, a language only has a few dozen phonemes. For instance, there are about 44 phonemes in English, with some variation introduced by dialects and accents. By recognizing phonemes rather than whole words, ASR systems are better able to handle variations in speech, including speech rate and pronunciation. They can also better cope with background noise and other challenging audio conditions. When output words are being dynamically assembled from phonemes, ASR systems can potentially recognize words that were not in their original training set, including new words, proper names, and other vocabulary. This, however, is limited by the contents of the ASR vocabulary. Also, phonemes are not the only way to deconstruct words. A *grapheme* is the smallest unit of writing that can represent a phoneme, for instance, the grapheme "ch" represents the phoneme [tʃ]. A *morpheme* is the smallest unit of meaning, it can be a word, a part of a word, or a prefix or suffix. A *lexical* morpheme carries the meaning of a word ("dog", "catch"). A *grammatical* morpheme does not have an inherent meaning, but it changes the meaning of a word ("-s" to make a noun plural, "-est" to make an adjective a superlative). In linguistics, a word can be expressed as a sequence of phonemes, a sequence of graphemes, a sequence of morphemes, a part-of-speech tag (noun, verb, adjective) with its corresponding flexional information (grammatical gender, declension), a syntactic unit (subject, object), or even as a semantic representation (word embedding). Modern ASR systems perceive words as sequences of phonemes, which makes the training process for ASR more feasible, flexible, and efficient.

Given the vocabulary W of words in a task or language, the goal of the ASR system is to find the sequence of words $W^* = (w_1, w_2, \ldots, w_n), \forall i : w_i \in \mathrm{W}$ such, that:

$$W^* = \arg\max_{\mathrm{W}} P(\mathrm{W}|Y) = \arg\max_{\mathrm{W}} \frac{P(Y|\mathrm{W})P(\mathrm{W})}{P(Y)} = \arg\max_{\mathrm{W}} P(Y|\mathrm{W})P(\mathrm{W}) \tag{4.2}$$

where Y is the set of observable acoustic features, and thus $P(Y)$ is the same for all possible sequences of words. The Equation 4.2 clearly demonstrates that the task of the ASR can be divided into two parts: finding the most probable acoustic features given what words have been spoken ($P(Y|\mathrm{W})$), and finding the most probable words ($P(\mathrm{W})$). The first task is known as *acoustic modeling*, and the second task is the *language modeling*. The process of speech recognition is presented in Fig. 4.1.

Interestingly, the statistical approach to language modeling allows to reverse the process of speech recognition and transform it into the process of speech synthesis. Starting from the language model $P(\mathrm{W})$ one can generate a sequence of words $W^* = (w_1, w_2 \ldots, w_n)$, and then transform W^* using a pronunciation mapping $\Pi : \mathrm{W} \rightarrow \mathrm{P}$ into a sequence of phonemes $\mathrm{P}^* = (\rho_1, \rho_2, \ldots, \rho_m)$. This sequence of phonemes can be passed through the acoustic model $P(Y|\mathrm{W})$ to produce the sequence of acoustic features $Y = (y_1, y_2, \ldots, y_k)$ which can be synthesized back into a speech signal.

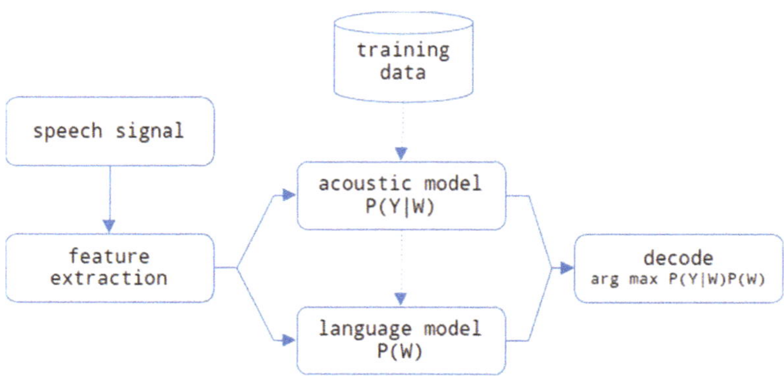

Fig. 4.1 Process of speech recognition

Feature Extraction from Speech

Feature extraction from speech presents significant challenges due to the complexity and variability of human speech patterns. This complexity arises from factors such as individual vocal characteristics, language nuances, emotions, and environmental noise, all of which can alter the speech signal. The non-stationary nature of the speech signal adds another layer of difficulty in feature extraction. Consequently, the task of accurately extracting and interpreting meaningful information from speech signals requires sophisticated techniques and remains an active area of research [14, 23]. Here we only touch upon these complexities without diving into the detailed aspects of this topic and describe the feature extraction from speech to the degree necessary for the purpose of this book.

When performing feature extraction from speech, we are seeking features with typically desirable characteristics for machine learning. We are interested in features that are relevant, robust (consistent and reliable under varying conditions, speakers, languages), computationally efficient (for real-time or near-real-time applications of speech processing), and discriminative. In particular, the robustness of features is important. We want features that are not influenced by the change in the fundamental frequency $F0$ or by the introduction of background noise and channel distortions. In addition, we want to avoid redundancy among features. If a feature f_i is derived from the glottal spectrum and correlates with the feature f_j extracted from the formants of the vocal tract, the inclusion of both f_i and f_j may not be desirable.

The input speech signal is non-stationary, meaning that it changes over time. Sampling is the process of converting a continuous-time signal into a discrete-time signal by taking measurements at regular intervals, known as the sampling rate. Typical sampling frequencies for speech processing are between 8 kHz and 48 kHz. The sampled signal is usually approximated by a finite set of values using the process of *quantization*. This involves converting a continuous-amplitude signal into a digital signal. A window of such samples of a fixed duration (usually 20-40 ms) is called a *frame*. The speech signal is assumed to be stationary within the frame, however, there may exist discontinuities at the beginning and end of each frame. To reduce the negative effects of those discontinuities, known as *spectral leakage* (if the frames are not long enough to capture the entire period of the speech signal, then the energy from the signal can leak into neighboring frames), the technique of windowing is applied. Windowing is the process of multiplying the frame with a window function, such as the Hamming window (also known as the raised cosine window). This function smoothes the edges of a signal before applying a Fourier transform by applying the following formula:

$$w(n) = 0.54 - 0.46 \cos\left(\frac{2\pi n}{N-1}\right) \tag{4.3}$$

where n is the sample index and N is the total number of samples in a frame. After windowing, each frame of speech can be analyzed individually, and features can be extracted.

The obvious choice of features is a set of statistical characteristics of the signal, such as mean, median, minimum, maximum, standard deviation, and quartiles of the frequencies of each signal. A less-known feature is the energy of the signal defined as

$$E(x) = \sum_{n} x(n)^2 \tag{4.4}$$

and the root mean square energy defined as

$$RMS_E(x) = \sqrt{\frac{1}{N} E(x)} \tag{4.5}$$

Probably the most common features specific to speech signal are the *mel frequency cepstral coefficients* (MFCC). The idea behind MFCC is to produce features that approximate the way the human ear (in particular, the cochlea) perceives sound. MFCCs are computed over short frames with a rolling window of 10 ms. Each frame is first subjected to the Discrete Fourier Transform (DFT):

$$S_j(k) = \sum_{j=1}^{N} s_j(n) w(n) e^{-i \frac{2\pi}{N} kn} \quad 1 \leq k \leq K \tag{4.6}$$

where $s_j(n)$ is the framed signal, $w(n)$ is the Hamming window, N is the number of samples in the Hamming window, and K is the number of harmonics in the DFT. Next, the periodogram estimate of the power spectrum is computed as:

$$P_j(k) = \frac{1}{N} S_j(k)^2 \tag{4.7}$$

The next step is to "clean" the periodogram and project it to the mel scale. The reason for this step is that the human ear is not good at discriminating between closely spaced frequencies, in particular for high frequencies. A simple solution is to cluster together similar frequencies into bins. The clustering is performed using the mel filterbank, a set of triangular-shaped filters of predefined widths and positions in the frequency amplitude space. Filters in the mel filterbank are narrow for low frequencies and grow wider as we move toward higher frequencies. By applying the filterbank to the periodogram power spectrum, we receive the filtered power spectrum with well-defined signals, i.e., we have the information on how much energy is carried in the speech signal around certain mel-spaced frequencies. The number of features is equal to the number of filters in the filterbank (usually between 20 and 40). Figure 4.2 presents an example of such a filterbank.

The human ear does not perceive the increase in loudness of the input signal linearly, but logarithmically. Thus, it is necessary to project the powers of spectrum onto the *mel scale* (melody scale). In the mel scale pitches are perceived by listeners to be equally spaced apart. The mel scale is established in parallel with the standard

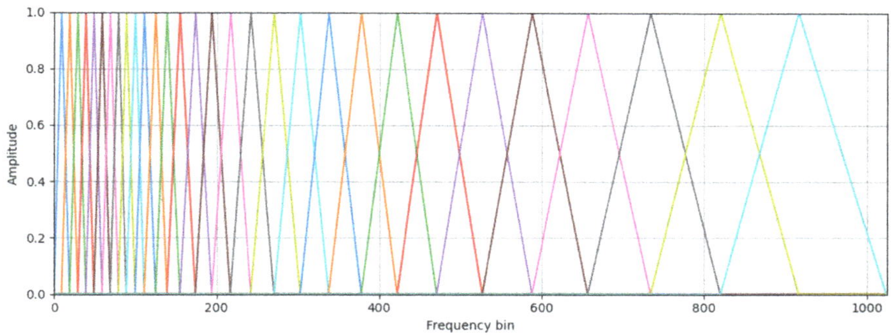

Fig. 4.2 Example of a mel filterbank with 30 triangular filters

frequency scale by equating 1000 mels to a 1000 Hz tone that is 40 dB above the listener's threshold. The formula for the conversion of frequency f in hertz to m in mels is the following:

$$m = 2595 \log_{10}(1 + \frac{f}{700}) = 1125 \ln(1 + \frac{f}{700}) \tag{4.8}$$

The final step is to perform the Discrete Cosine Transform [1] to the logarithm of energy contained in each filter:

$$DCT[k] = \sum_{i=0}^{N-1} \log E_i \cos\left(\frac{\pi}{N}\left(i + \frac{1}{2}\right)k\right), \quad k = 0, \ldots, N-1 \tag{4.9}$$

where k is the index of the frequency component, i is the index of the filterbank, and E_i is the energy in filter i. The Discrete Cosine Transform de-correlates the energies (which are typically highly correlated) and yields a set of mel frequency cepstral coefficients (MFCCs). The number of coefficients extracted as features per frame varies depending on the application. For speech recognition, usually, the lowest 10–15 coefficients are kept.

The MFCCs represent the power spectral envelope of each frame, but they do not convey information about the dynamics between frames. Obviously, speech signal changes over time and this change should also by quantified using some features. A popular choice to describe the dynamics of MFCCs is the computation of Δ and $\Delta\Delta$ (deltas, and delta-deltas). The formula to compute Δ is as follows:

$$\Delta(f) = \frac{\sum_{n=1}^{N} n(c_{f+n} - c_{f-n})}{2\sum_{n=1}^{N} n^2}$$

$$\Delta(f) = \frac{\sum_{n=1}^{N} n(c_{f+n} - c_{f-n})}{2\sum_{n=1}^{N} n^2} \tag{4.10}$$

where f is the frame, c_f is the value of the mel frequency cepstrum coefficient for that frame, and N is a constant (usually $N = 2$). The formula for $\Delta\Delta$ is identical with the only difference being the fact, that the computation is not performed on static coefficients, but on the deltas. Both deltas and delta-deltas (sometimes referred to as mel frequency cepstral differential coefficients) are added as features to each frame. Their inclusion consistently improves the accuracy of ASR systems on a wide variety of tasks.

A *spectrogram* is a visual representation of the spectrum of frequencies in a speech signal as they vary with time. It is essentially a 2D graph with three dimensions: time, frequency, and amplitude. Time is represented on the horizontal axis, frequency on the vertical axis, and the amplitude of a given frequency at a specific time is represented by the intensity of color or brightness at that point in the graph. Many features can be extracted directly from a spectrogram: spectral centroid, spectral bandwidth, spectral contrast, and spectral roll-off, to name a few. In addition to these, time-based features such as zero-crossing rate and temporal centroid, and perceptual features like loudness and sharpness can also be derived from spectrograms. These features are more useful in tasks such as music information retrieval, audio classification, and sound synthesis. Figure 4.3 presents the spectrogram of the first 60 seconds of the *Imperial March* from the movie "Star Wars".

The final three features describing speech are jitter, shimmer, and noise-to-harmonic ratio. These features are used mostly in voice disorder diagnosis, speaker recognition, and emotion detection, and they do not contribute significantly to speech recognition. *Jitter* measures the cycle-to-cycle variations in the fundamental frequency $F0$. This variation is usually caused by the physiological changes or irregularities in the vocal fold vibration during speech production. High levels of jitter typically indicate a rough or hoarse voice, often associated with vocal pathologies.

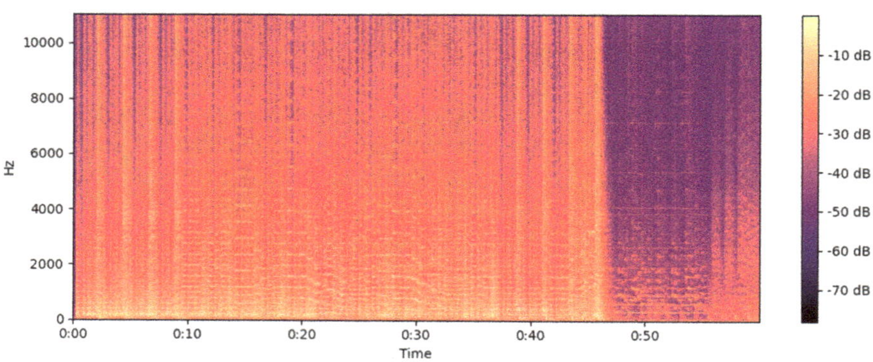

Fig. 4.3 Example of a spectrogram

However, small amounts of jitter are natural and contribute to the unique characteristics of each individual's voice. *Shimmer* is a very similar feature that measures the cycle-to-cycle variation in speech amplitude. An increased level of shimmer can indicate breathy voice quality, which may be indicative of voice disorders. Like jitter, shimmer is typically measured in percentages. The *Noise-to-Harmonic Ratio* (NHR) measures the proportion of the signal's energy that is attributed to noise as compared to the harmonic or periodic components of the speech signal. This feature is particularly useful in identifying irregularities in the voice signal, as an increased NHR value may indicate a higher presence of noise, suggesting possible voice disorders.

Acoustic Modeling with HMM-GMM

Acoustic modeling plays a key role in the development and functioning of ASR systems. Its primary function is to transform speech signals into a sequence of linguistic units. Among the various techniques used for acoustic modeling, the Hidden Markov Model-Gaussian Mixture Model (HMM-GMM) approach stands out as one of the most widely used and extensively studied methods. Its popularity can be attributed to its ability to effectively capture the temporal dynamics of speech signals and its flexibility in modeling complex acoustic patterns. As a result, HMM-GMM has been successfully applied in numerous ASRs, making it a go-to method for acoustic modeling. Although deep neural networks have been gaining prominence in recent years, HMMs are still relevant for speech recognition due to their simplicity, speed, and accuracy.

As we have said, the main task of the acoustic model is to transform a sequence of audio features into a sequence of phonemes. HMMs are a natural candidate for modeling sequences, so let us briefly discuss this technique. HMMs are statistical models used in machine learning and pattern recognition to transform one sequence into another. In a HMM, the sequence is represented as a chain of states, where each state corresponds to an element or a segment of the sequence. The transitions between states represent the probability of transitioning from one element to another in the sequence. The key concept in HMMs is that the states are not directly observable, hence the name "hidden." Instead, each state emits an output (observation) according to a probability distribution associated with that state. These emitted observations provide indirect information about the underlying state.

Let us consider an example of using HMMs to transform a sequence of audio features into a sequence of phonemes. In this case, the states in the HMM would correspond to the phonemes, and the transitions would represent the likelihood of transitioning from one phoneme to another. The audio features, such as MFCC coefficients or spectrograms, serve as the observations emitted by each state. By training the HMM on a dataset of audio recordings and their corresponding phoneme sequences, the model learns the probabilities of transitioning between phonemes and the probability distributions over the audio features for each phoneme. As the result, the model learns the structure of spoken language. For instance, in English, the pair of phonemes /s/ and /t/ is highly probable. This pair is often found in words like "sit," "stop," or "fast." Phonetically, the /s/ sound is an unvoiced alveolar fricative produced with the tip of the tongue against the alveolar ridge, while /t/ is an unvoiced alveolar stop produced by briefly stopping the airflow behind the alveolar ridge before releasing it. On the other hand, the pair of phonemes /p/ and /h/ is highly improbable when they occur consecutively. The /p/ sound is an unvoiced bilabial stop produced by closing the lips and releasing a burst of air, while /h/ is a voiced glottal fricative produced by exhaling air through a partially constricted glottis. These two sounds don't naturally occur in consecutive order within English words, making this pair highly unlikely. So, the training of the HMM provides data to compute the probabilities of transitions between states. During the decoding or inference phase, given a sequence of audio features, the HMM can determine the most likely sequence of phonemes that generated those features. This process involves estimating the probabilities of different phoneme sequences using the observed features and the transition probabilities learned during training.

For speech recognition the HMM is slightly modified, because a phoneme is not discrete, its expression can span over several frames of audio features. Thus, in order to capture longer phonemes that span over many frames, each phoneme is modeled as a 3-state (or sometimes 5-state) linear HMM. Conceptually, one can think of 3-state HMM as a model to recognize the beginning, the middle, and the end of each phoneme. Since HMMs are context-independent with respect to observations, the probability of observing the phoneme ϕ at time t depends only on the current state s_t and not on previous states:

$$p(o_t = \phi | s_t, s_{t-1}, \ldots, s_1) = p(o_t = \phi | s_t)$$

$$p(o_t = \phi | s_t, s_{t-1}, \ldots, s_1) = p(o_t = \phi | s_t) \tag{4.11}$$

This assumption of independence between phoneme states is quite rigid. It can be slightly weakened by turning to context dependent phonemes, known as *triphones*. A triphone is a sequence of three phonemes, where the middle phone represents the target sound of interest, while the phones on either side (known as left and right context phones) provide context for the target sound. An example of a triphone is /b-ay-t/ which represents the sound "bite", or /ʃ-o-p/ which represents the sound "shop". Triphones capture the contextual information of a phoneme by considering the influence of neighboring phonemes, and allow for better handling of coarticulation and phonetic variations that occur due to adjacent sounds. This enhances the discrimination between similar sounds and helps improve the accuracy of ASR systems. Of course, there is a price to pay: the model size increases and the inference time can become longer when compared to single phoneme modeling.

The second part of the acoustic model mentions the use of Gaussian Mixture Models (GMM). GMMs provide a flexible way to model probability distributions by combining multiple Gaussian distributions. Each Gaussian component represents a cluster of observations in the feature space. In the context of acoustic modeling in ASR, GMMs are used to model the acoustic characteristics of phonemes or phonetic units. The goal is to estimate the probability distribution of acoustic feature vectors for each phoneme. In acoustic modeling, each state of the HMM is associated with a specific phoneme or sub-phoneme, and a GMM is trained to model the acoustic distribution of that state. The GMM parameters, which include the means, variances, and weights of the Gaussian components, are estimated using a training dataset. During recognition, the GMM for each state is used to calculate the likelihood of the observed acoustic features, and the HMM framework combines these likelihoods to decode the speech and determine the most likely sequence of phonemes. In other words, the GMM predicts the probability of an input feature being generated from the current state. The combination of GMMs and HMMs allows for robust modeling of the acoustic properties of speech. The GMMs capture the variability within each phonetic unit, accounting for different pronunciations and speaker-dependent characteristics, while the HMMs handle the temporal aspect of speech, capturing the transitions between different phonetic units.

Finally, let us briefly discuss the training procedure of acoustic models based on the HMM-GMM framework. The training involves two key steps: training the HMM parameters using the Baum-Welch algorithm (also known as the Forward-Backward algorithm) and training the GMM parameters using the Expectation-Maximization (EM) algorithm. The forward pass of the Baum-Welch algorithm computes the forward probabilities, which are the probabilities of being in a particular state at a specific time given the previous observations. It involves recursive calculation of the probability of transitioning from one state to another and the emission probability of the observed feature vector. The backward pass computes the backward probabilities, which are the probabilities of being in a particular state at a specific time given the future observations. The combination of forward and backward probabilities is used to calculate the state posterior probabilities and transition probabilities needed for updating the HMM parameters. The Baum-Welch algorithm performs multiple iterations of the forward-backward passes, updating the HMM parameters until convergence. For each state of the HMM the EM algorithm trains the parameters of the GMM for that state. In the expectation step, given the current estimate of the GMM parameters, the probabilities of each Gaussian component generating the observations are computed. This is done using the Gaussian probability density function. In the maximization step, the GMM parameters (means, variances, and weights) are updated by maximizing the log-likelihood of the observed data. This is done by using techniques like Maximum Likelihood Estimation (MLE) or Maximum A Posteriori (MAP) estimation. The expectation and maximization steps are repeated iteratively until convergence, ensuring the GMM parameters are adjusted to better recognize the individual observations.

Acoustic Modeling with HMM-DNN

In recent years, there has been a shift from using Hidden Markov Model-Gaussian Mixture Model (HMM-GMM) acoustic models towards using Deep Neural Networks (DNNs) for acoustic modeling in speech recognition [24]. This shift is driven by the superior performance of DNNs in capturing complex and non-linear relationships in speech data. DNNs are neural networks with multiple hidden layers that can learn hierarchical representations of data. They have shown remarkable success in various domains, including image recognition and natural language processing. In the context of speech recognition, DNNs have demonstrated superior performance in capturing acoustic patterns and handling large-scale vocabulary tasks. When used in conjunction with HMMs, DNNs are typically employed as acoustic models to predict sequences of phonemes from raw audio features. This architecture is often referred to as a Hybrid HMM-DNN system. The raw audio features, such as MFCCs or filterbanks, are fed into the DNN as input. The DNN learns to map these acoustic features to the posterior probabilities of phonemes. When compared to HMM-GMM, DNN predicts the probability of observing a sequence of phonemes given the input rather than describing the likelihood of the input given the current state. In order to achieve satisfactory results, the number of hidden layers in a DNN is large. Audio features are extracted from much larger windows in order to provide more context and discount variations in individual pronunciations of phonemes. The training corpora for HMM-DNN must be also much larger to allow the neural network to train.

During training, the DNN is trained to minimize the difference between the predicted posterior probabilities and the target phoneme alignments, which are obtained either through forced alignment or by using manual transcriptions. The HMM framework then combines the sequence of posterior probabilities predicted by the DNN with the language model probabilities to generate the most likely sequence of words or phonemes. This integration of DNNs and HMMs offers several advantages. DNNs provide better modeling capabilities for capturing complex patterns and long-term dependencies in the speech data. HMMs, on the other hand, handle the temporal aspect of speech and provide a framework for decoding and modeling the sequential nature of speech. Together, DNNs and HMMs have demonstrated significant improvements in speech recognition accuracy.

End-to-end Acoustic Models

The concept behind end-to-end acoustic modeling [20] in ASR is to directly map the input audio signal to the corresponding text outputs, eliminating the need for explicitly designing intermediate representations or modeling steps. In end-to-end ASR, the entire acoustic-to-text mapping is learned by a single deep neural network, simplifying the ASR pipeline. To achieve end-to-end acoustic modeling, deep network architectures that combine feature extraction and phoneme prediction steps are used. One popular architecture is the Connectionist Temporal Classification (CTC) model [19, 33], which incorporates recurrent neural networks (RNNs) or their variations such as Long Short-Term Memory (LSTM) [25] or Gated Recurrent Units (GRUs) [8]. In the CTC-based architecture, the deep network directly takes the raw audio waveform as input and learns to produce character or phoneme-level predictions. The network learns to align the input waveform with the corresponding transcript without the need for explicit alignment information. Through the use of recurrent connections, the network models the temporal dependencies and uses contextual information from the entire utterance. Another approach is the Listen, Attend and Spell (LAS) architecture [6], which combines an encoder-decoder framework with attention mechanisms. The encoder network processes the input audio spectrogram to obtain a higher-level, context-rich representation. The decoder network then generates the sequence of characters or phonemes while attending to relevant parts of the encoder's output through the attention mechanism. End-to-end acoustic modeling in ASR has gained attention due to its simplicity and promising results in various speech recognition tasks, even though it may require larger amounts of training data compared to traditional approaches.

An example of a modern large-scale end-to-end acoustic model is Whisper developed by OpenAI [47]. Whisper utilizes a deep learning architecture to directly convert raw audio waveforms into text transcriptions, eliminating the need for intermediate steps in the speech recognition pipeline. It is trained on a massive dataset consisting of 680 000 hours of multilingual and multitask supervised data collected from the Web. 117 000 hours cover 96 languages other than English, with additional 125 000 hours of translations from a source language to English. The main reason why such a diverse dataset has been collected is the observation, that speech recognition systems that are pre-trained on a number of datasets with supervision are more robust and generalize to held-out datasets much better than models trained on a single dataset. This is true even in presence of low quality datasets, although the authors of Whisper included many cleaning tasks in the pre-processing pipeline, trying to remove transcripts that were generated automatically (e.g. subtitles of YouTube videos). This extensive training corpus allows Whisper to learn a wide range of acoustic and linguistic patterns, enabling it to perform well in various languages and tasks.

The main problem addressed by Whisper is the gap between large-scale unsupervised training of ASRs (1 000 000 hours of training data) and supervised training (thousands of hours of training data). The pre-trained audio encoders learn high quality speech representation, but lack tools for efficient decoding to usable outputs. Small supervised training datasets are too narrow and overfit to the training dataset, resulting in sub-par performance across different datasets. The solution is two-fold: compile a very large collection of human-transcribed audio, and train a universal speech recognition system in a supervised way.

Whisper incorporates a new self-supervised objective called contrastive predictive coding, where it learns to predict future frames of the audio signal. This self-supervised training scheme helps the model capture rich representations of the input audio. It also utilizes advanced training techniques like curriculum learning and large-scale distributed training on thousands of GPUs to further enhance its performance. The architecture of the model is an encoder-decoder Transformer trained on Mel spectrograms with 25 ms window and 10 ms stride. The model is trained for multitasking (language identification, voice activity detection, translation, time-aligned transcription, text-only transcription) for 2^{20} updates (2-3 epochs).

The Whisper model achieves impressive results, outperforming traditional hybrid systems and previous end-to-end architectures. It demonstrates high accuracy in speech recognition tasks, showcasing the potential of large-scale end-to-end models in transforming the field of speech recognition. With its ability to handle multiple languages and tasks, Whisper serves as an example of a state-of-the-art end-to-end acoustic model that pushes the boundaries of large-scale deep learning in the domain of speech recognition. Across 13 different datasets, the average word error rate (WER) achieved by the large Whisper model is 12.9%. In terms of multi-lingual speech recognition, Whisper has achieved state-of-the-art results, although it falls behind the XLS-R and mSLAM models on the VoxPopuli dataset. When it comes to translation, Whisper excels in low- and mid-resourced languages, but mSLAM outperforms it in high-resourced languages. Regarding language identification, Whisper does not perform well, however, it is worth noting that it has no training data available for 20 out of the 102 languages present in the benchmark Fleur dataset. In the field of long-form transcription, Whisper demonstrates comparable performance to commercial ASRs. For zero-shot learning, Whisper shows remarkable robustness, coming close to human levels when compared to models trained on a single dataset. Furthermore, Whisper exhibits strong resilience to white and pub noise, maintaining robustness up to 10dB.

The original Whisper model has been updated twice. The Whisper v2 model was published on December 8, 2022. This version, referred to as the `large-v2 model`, was trained for 2.5 times more epochs than its predecessor and included enhancements such as SpecAugment and other regularization techniques, leading to improved performance in transcription, translation, and language identification tasks compared to the original `large-v1` model. The latest version of the model, referred to as `large-v3`, was announced by OpenAI on November 6, 2023. Whisper `large-v3` maintains the architecture of its predecessors with a few minor adjustments: it uses 128 Mel frequency bins in the spectrogram input instead of 80 and includes a new language token for Cantonese. This model was trained on a dataset consisting of 1 million hours of weakly labeled audio and 4 million hours of pseudo-labeled audio collected using Whisper `large-v2`. As a result, Whisper `large-v3` demonstrates a 10% to 20% reduction in error rates compared to its predecessor across various languages. However, some users have reported that in their testing, Whisper `large-v2` still outperforms `large-v3` by 20-30% on average, with the latter exhibiting more hallucinations. The reasons behind this discrepancy are still under investigation.

Another model that has emerged recently in the field of speech recognition and translation is Canary [46]. This open-source encoder-decoder model supports English, French, Spanish, and German languages. Unlike its counterparts, Whisper and SeamlessM4T, Canary achieves high accuracy with only 86 000 hours of training data, a fraction of what other models require. Canary's architecture uses a FastConformer-based attention encoder-decoder [50], which is more efficient and faster compared to the traditional Transformer architecture used by Whisper. The model employs several techniques to enhance data efficiency: synthetic data generation for AST, data balancing, dynamic data blending, and noise-robust fine-tuning. These methods help Canary use its smaller dataset effectively. Training Canary takes less than 48 hours on 128 NVIDIA A100 80GB GPUs, significantly faster than Whisper's training time. Despite requiring less data and training time, Canary outperforms Whisper and similar models on various ASR and AST benchmarks.

4.3.3 Pronunciation Modeling

Let us get back to Eq. 4.2 where we have defined the goal of the ASR as finding a sequence of words $W^* = \arg\max P(Y|W)P(W)$. When the vocabulary of the ASR is limited to a very specific domain, it may be possible to collect enough examples of each word to successfully train the model which translates from audio features to words. In practice, however, this is not the case. In large vocabulary tasks (i.e. when the output of the ASR is not constrained to a narrowly defined domain) most words will almost certainly not be collected at all, and the majority of words will be represented by at most 1-2 examples in the training dataset. Thus, instead of predicting words, we have to predict phonemes, and then combine phonemes into words. This, however, introduces a new problem: pronunciation is highly idiosyncratic, varying among individuals and even influenced by temporary factors such as speaking style, emotion, or health. Additionally, different accents and dialects can further contribute to the variability in the way words are pronounced. As a result, a single word can have multiple alternative phonemic transcriptions. To address this issue, it becomes necessary to introduce a specialized pronunciation model within the ASR system. This model acts as a mapping mechanism that can accurately combine a given sequence of phonemes into words, taking into account the specific pronunciation variations and linguistic context. We modify the original definition of the goal of ASR by assuming that words are sequences of phonemes:

$$W^* = \arg\max_{\mathrm{W}} P(Y|\mathrm{W})P(\mathrm{W}) \tag{4.12}$$

$$= \arg\max_{\mathrm{W}} P(Y|Z, \mathrm{W})P(Z, \mathrm{W}) \tag{4.13}$$

$$\approx \arg\max_{\mathrm{W}} P(Y|Z) \sum_{Z} P(Z|\mathrm{W})P(\mathrm{W}) \tag{4.14}$$

$$\approx \arg\max_{\mathrm{W}} \max_{Z} P(Y|Z)P(Z|\mathrm{W})P(\mathrm{W}) \tag{4.15}$$

Now we are asking for a sequence of words W^* which maximizes the probability of observed audio features given the pronunciation ($P(Y|Z)$) multiplied by the probability of the pronunciation given the words ($P(Z|W)$), and multiplied by the language probability of words ($P(W)$). The element $P(Z|W)$ is called the *pronunciation model* and it represents the assessment of the variability of how a given sequence of words can be produced.

The simplest form of pronunciation models are lexicons hand-crafted by experts and mapping words into sequences of phonemes. This process requires a detailed understanding of the phonetic structure of the language. Linguists often construct a linguistic tree, organizing words based on similarities in their pronunciation patterns. This tree serves as a guide to identify common phonetic features and group words accordingly. For instance, one branch of the tree may include words like "cat" / kæt/, "sat" / sæt/, and "hat" / hæt/ that share similar vowel sounds and end in the same consonant sound. Another branch may contain words like "either" / iːðər/ and "neither" / naɪðər/ that demonstrate the pronunciation of very similar words. The construction of lexicons is further complicated by alternative pronunciations of the same word (e.g., the word "schedule" can be pronounced as / ˈʃɛd.jul/ (sh ɛd-yool) or /skɛd.jul/ (sk ɛd-yool). On the other hand, there are multiple examples of words with different spelling and meaning that share the same IPA transcription, for instance "hear" and "here" (/hɪər/), "right" and "write" (/raɪt/), or "maid" and "made" (/ meɪd/). Unfortunately, the problem of sufficient coverage remains to be solved. Also, pronunciation lexicons do not contain probabilities for phoneme sequences as these are extremely difficult to evaluate manually.

The construction of the pronunciation model requires the definition of the words that will be included in the final dictionary. Obviously, the selection of dictionary words should aim at the maximization of the coverage of the test set (or, more generally, the coverage of words that the model will encounter after deployment). Each word included in the dictionary should be represented by an entry in the pronunciation model. It is impossible to guarantee the full coverage of the domain and the presence of *out-of-vocabulary* (OOV) words is certain. The *OOV-rate* is the percentage of words in the test set that were not included in the ASR dictionary. Careful construction of dictionaries for ASR is very important and requires domain expertise, because there is a direct relationship between the OOV-rate of the system and its accuracy. Additional challenge arises in case of inflective languages, where minor modifications of a word denote its grammatical gender, tense, case, voice, number, etc. Representing all inflections of a single word in the ASR dictionary is impractical. To address this problem, the ASR dictionary can be constructed to contain subword units, such as morphemes or even individual characters. By including subword units, ASR systems can handle OOV words more effectively. For example, instead of having an OOV word that has never been seen before, the system can potentially break it down into familiar morphemes or characters to provide a more accurate transcription. Additionally, ASR dictionaries can also incorporate frequent sequences of words by using bigrams or n-grams. These sequences of words often occur together with high probability in natural language. By including these common word sequences, the ASR system can improve the accuracy of transcriptions for such collocations or multi-word expressions.

Finally, the construction of the ASR dictionary requires the design of the normalization schema. Word normalization helps ensure accurate and consistent transcriptions by converting different variations of words into a standardized form. Proper names, abbreviations, geographical names, and other specialized terms often pose challenges due to variations in spelling. Normalization allows these variations to be mapped to a unified representation in the dictionary. For example, abbreviations like "Dr." or "Dr" can be normalized to "dr" to ensure consistency. Geographical names, such as "New York", "N.Y.C" or "NYC," can be normalized to their standardized forms. Normalization also applies to digits and numbers. For instance, "100" can be transformed into "one hundred" for better readability. This normalization process extends to handling specific numerical formats, dates, times, currencies, and other numerical expressions. ASR systems can utilize external knowledge bases or resources. These knowledge bases, such as Wikipedia or domain-specific databases, can provide information on canonical forms or alternative spellings.

One interesting alternative to explicit pronunciation modeling using lexicons is the *grapheme-to-phoneme* (G2P) approach [13, 48]. G2P models are language models that aim to predict the pronunciation of a word based on its written form, mapping graphemes (letters or letter combinations) to phonemes. These models can assist in predicting the phonetic representation of a word, which is important for training acoustic models that map audio features to phonemes. By accurately predicting the phonemes of a word, G2P models can enhance the performance of acoustic models by providing improved alignments between audio and phonetic units, and provide probability distributions over different pronunciations of words. The graphemes can be also used directly to remove the need of pronunciation modeling in acoustic models. Recent works have shown that graphemes are useful for multi-accent pronunciation modeling [49] as well as for multi-lingual pronunciation modeling [28]. Graphemes can also be used in an end-to-end setting when the translation proceeds directly from audio features to characters or other morphemes. One such model is wav2letter [12] designed to operate solely on raw audio waveforms, without relying on intermediate representations such as spectrograms or Mel-frequency cepstral coefficients. By directly modeling the relationship between graphemes and audio, wav2letter can effectively transcribe speech into text, even for languages with complex or variable pronunciation.

4.3.4 Language Modeling

Let us recall the Eq. 4.2 describing the entire process of speech recognition:

$$W^* = \arg\max_{\mathrm{W}} P(\mathrm{W}|Y) = \arg\max_{\mathrm{W}} \frac{P(Y|\mathrm{W})P(\mathrm{W})}{P(Y)} = \arg\max_{\mathrm{W}} P(Y|\mathrm{W})P(\mathrm{W}) \tag{4.16}$$

In this equation the term $P(\mathrm{W})$ represents the *language model* which is tasked with assigning the probability to a given sequence of words. The need for a language model stems from the fact, that for a given sequence of phonemes there can be multiple mappings of phonemes to words. It is far more likely that the transcript contains the phrase "*my iphone is running out of battery*" than "*my eye foam is ruining our bakery*

". Importantly, the language model estimates the probability of the entire sequence of words and can overcome predictions resulting from a specific set of audio features (e.g., accent, unclear pronunciation, background noise injection). Even if a particular sequence of words is strongly suggested by the acoustic model (for instance, "*first of man*"), the language model can evaluate an alternative hypothesis ("*first of May*") as more likely to appear in the general language. We may think of the language model as the normalization of the variability introduced by the acoustic model.

The most obvious way to compute $P(W)$ is to use the relative frequencies:

$$P(w_n|w_1, w_2, \dots, w_{n-1}) = \frac{|w_1, \dots, w_i|}{\sum_v |w_1, \dots, w_{i-1}, v|} \tag{4.17}$$

where the number of occurrence of a sequence of words is computed over some large language corpus. The probabilities of sequences of words can be computed using chaining:

$$P(w_1, w_2, \dots, w_i) = P(w_1)\Pi_{k=2}^{i} P(w_k|w_1, \dots, w_{k-1}) = \Pi_{k=1}^{i} P(w_k|w_{k-1}) \tag{4.18}$$

Unfortunately, this approach works only for very short sequences. As the number of words in a sequence increases, the probability of encountering a pair of words such that $P(w_l|w_{l-1}) = 0$ grows. Let us review the options for mitigating this problem.

Smoothing

In language modeling, smoothing techniques are employed to address the issue of zero probabilities when encountering unseen or rare sequences of words in the training data. Smoothing helps adjust the probability estimates by redistributing some probability mass from seen sequences to unseen ones. Two common smoothing methods are Laplace smoothing (add-one smoothing) and add-k smoothing.

Laplace smoothing is a simple technique where a constant value (usually 1) is added to the count of each sequence. This smooths the probability distribution and ensures that no probability estimate is zero. Add-k smoothing is a generalization of Laplace smoothing, where instead of adding 1, a constant value k is added to each count. This allows for more flexibility in controlling the amount of smoothing applied. For instance, if we set k to 0.5, it would have a less pronounced effect compared to Laplace smoothing. The choice of the smoothing constant can significantly impact the probability estimates. Both Laplace and add-k smoothing techniques provide a simple and intuitive way to handle unseen or rare sequences in language modeling. However, it's important to note that while smoothing mitigates the problem of zero probabilities, it introduces a form of bias towards unseen events, as probabilities are shifted from known sequences to unseen ones.

Context-free Grammar

The language model can be based entirely on context-free grammar, which relies on a set of production rules that dictate the structure and composition of valid sentences in a language. In this model, speech recognition is limited to extreme cases where the speaker's vocabulary is highly constrained. The key aspect of this model is that recognition is binary, meaning it determines whether the input aligns with the grammar or not, without considering more nuanced variations or context.

The application of this language model for speech recognition involves defining a grammar with specific rules and production patterns. Each word or phrase in the limited vocabulary is assigned a corresponding terminal symbol, while the grammar rules outline the valid combinations and order of these symbols within sentences. The recognition process involves matching the input speech against the defined grammar and determining if the sentence adheres to the grammar rules. If the input aligns perfectly with the grammar, it is recognized; otherwise, it is rejected.

It is important to highlight that using a context-free grammar for speech recognition has significant limitations. The lack of flexibility in context-free grammars means they cannot handle variations, ambiguities, or understanding of natural language beyond what is explicitly defined in the grammar rules. This approach also restricts the vocabulary and structural complexity that can be recognized accurately. In practice, even limited use of context-free grammar for cases where precision of recognition is important (e.g. parsing of dates) may not be feasible due to the lack of sufficient flexibility. Therefore, while context-free grammar can be employed in very specific cases with strictly limited vocabularies, it falls short in addressing the vast range of language variations and natural speech encountered in most real-world scenarios.

N-gram Models

N-gram language models form the foundation of many NLP tasks, including speech recognition. The general idea behind n-gram models is to capture the probability distribution of sequences of words or phonemes in a language. By observing and analyzing a large corpus of training data, the models estimate the likelihood of encountering specific n-gram sequences (usually unigrams, bigrams, and trigrams). By calculating the probabilities of various n-grams, speech recognition systems can improve accuracy by selecting the most likely sequence of words or phonemes based on the language patterns learned from the training data. The idea is very simple, instead of trying to calculate the probability of the entire sequence of words, the n-gram model truncates the sequence to the last n elements:

$$P(w_1, w_2, \ldots, w_i) \approx P(w_{i-n+1}, w_{i-n+2}, \ldots, w_i) \tag{4.19}$$

N-gram models are a natural choice to implement an alternative technique to smoothing called *backoff*. Backoff is a technique used to estimate the probability of an n-gram by "backing off" to a lower-order n-gram model when the higher-order model has insufficient data. It allows for the smoothing of n-grams that have low frequencies or have not been observed in the training corpus. The intuition behind backoff is that if a higher-order n-gram has zero or few occurrences, we can rely on a lower-order n-gram model as a fallback option to estimate the probability. For example, suppose we have a trigram language model, but a specific trigram like "black crow dreams" has low frequency or is unseen. Instead of assigning it a zero probability, we can "back off" to a bigram model and estimate its probability using the underpinning bigram "crow dreams". If the bigram is also unavailable, we can further back off to a unigram model and approximate the probability with the unigram "dreams". Backoff is often implemented in a hierarchical manner, where the higher-order model is checked first, and if it fails, the model falls back to lower-order models until a non-zero probability can be computed. Each lower-order model contributes less context, but it ensures that probabilities are assigned even when specific n-grams are sparse. To insure less biased estimation of shortened sequence probability, an additional parameter is introduced (for the sake of readability the equation below assumes a trigram model):

$$P(w_i, w_{i+1}, w_{i+2}) \approx P(w_{i+1}, w_{i+2}) \cdot \beta(w_i, w_{i+1}) \tag{4.20}$$

where β controls the influence of the missed "part" of the higher-order n-gram:

$$\beta(w_i, w_{i+1}) = \frac{1 - \sum_{w_i, w_{i+1}, w_j \in C} P(w_i, w_{i+1} w_j}{1 - \sum_{w_i, w_{i+1} w_j \in C} P(w_{i+1} w_j)} \tag{4.21}$$

where C is the training corpus. By iterating only over n-grams appearing in the corpus we guarantee that the required probabilities are available.

Using n-gram models for speech recognition comes with both advantages and disadvantages. N-gram models tend to be computationally efficient during recognition, especially for smaller values of n. They also require less memory and processing power compared to more complex language models. This is an important advantage as time constraints in speech recognition are strict. Recall that the speech recognition is auto-regressive in the sense that a sequence of words recognized at timestamp t_n provides an important context for words to be recognized at timestamp t_{n+1}, so the inference of acoustic, pronunciation, and language models combined must be efficient. Another advantage of n-gram models is that they are easy to train on large corpora (their training amounts to efficient counting of n-gram instances in the training corpus). The main disadvantage of n-gram models is that the are inherently incapable of using long-range dependencies for context. Any information beyond the n-gram order is unavailable to the model and can significantly limit the predictive power. This is particularly relevant to spontaneous speech recognition where speaker utterances are constantly interrupted by back-channeling and other artifacts of unscripted dialog. Another disadvantage of n-gram models is data sparsity for higher n-gram orders. It is very unlikely that the training corpus will provide evidence for all possible n-grams as the number of observed n-grams decreases with the increase of the n-gram order. This leads inadvertently to backoff, or custom handling of out-of-vocabulary or unseen sequences.

Extensions to n-gram Language Models

Over the years, many extensions have been proposed to the basic n-gram language model scheme. Factored language models [3] replace each word in a sequence by a feature vector. Possible features include, but are not limited to, part of speech information, grammatical role in a sentence, or semantics. This improves the accuracy of word prediction and allows to utilize implicit dependencies between features. In other words, when performing backoff, the model is not limited to the probability distribution of sequences of words, but may use the probability distribution of a word given a history of part of speech. This can be useful to distinguish between homophones such as "*I want to write a letter*" and "*I want a right ladder*". Another interesting modification of the n-gram model is structured language modeling [7]. According to this proposal, n-grams are not extracted from the raw text, but from dependency trees. For instance, after performing syntax parsing of the sentence "*The man I saw yesterday was walking barefoot*", one of the 4-grams would be (*the*, *man*, *was*, *walking*) as "*the man*" is the nominal subject of the verb "*was walking*".

Cache language models [44] are an enhancement to traditional language models that aim to improve efficiency and computational speed without compromising accuracy. The idea behind cache language model is to use a cache, a memory component that stores recently encountered words or phrases, to speed up the language modeling process. In a cache language model, when predicting the next word in a sequence, the model first looks into the cache. If the current context matches a previously encountered context in the cache, the model directly retrieves the probability distribution associated with that context from the cache. If the current context is not found in the cache, the model relies on the traditional language modeling techniques to estimate the probability distribution of the next word. Cache language models strike a balance between efficiency and accuracy by utilizing a trade-off between the cache size and computational overhead. With a larger cache, the model has a higher chance of finding matching contexts and speeding up predictions. However, a larger cache requires more memory and computation to maintain. Therefore, cache language models often employ strategies to manage the cache size effectively, such as prioritizing more recent contexts or employing cache replacement policies.

Trigger language models [29] are specialized language models designed to predict specific trigger words or phrases in a given context. Unlike traditional language models that focus on predicting the next word in a sequence, trigger language models aim to identify specific events or conditions by detecting trigger patterns. Trigger language models are commonly used in applications such as information extraction, event detection, or sentiment analysis, where specific triggers play a key role in determining the occurrence of an event or the sentiment of a text. In speech recognition, trigger language models can be trained to identify trigger words or phrases associated with specific events, commands, or prompts that are frequently encountered in the application domain. These triggers can be used to initiate actions or guide the speech recognition system's behavior in contextually appropriate ways. For example, in a voice-controlled assistant application, a trigger language model can be trained to recognize trigger phrases like "Hey, assistant" or "Okay, assistant" to activate and listen for subsequent user commands. When such a trigger is detected, the system shifts to a specific mode for handling voice commands rather than regular speech recognition.

Another viable alternative for n-gram models are topic language models [17]. These models are trained on large corpora of text data where documents are assigned to specific topics. By understanding the underlying topics of the speech input, topic language models can create more meaningful and accurate predictions. In speech recognition, topic language models can be used to improve transcription accuracy by specializing in different topic domains. These models can capture the specific vocabulary, phrasing, and language patterns associated with each topic. For example, a speech recognition system used in a medical context can utilize a topic language model that is trained specifically on medical literature and terminology. This allows the model to handle domain-specific variants, acronyms and technical terms that may not be well-represented in the generic n-gram model.

Class-based n-gram models [4] are an extension of traditional n-gram language models that group words into classes based on their similarities in terms of their linguistic properties or semantic meaning. Instead of treating every word as a distinct unit, class-based n-gram models use a smaller set of classes to represent words, reducing the vocabulary size and alleviating the data sparsity problem faced by traditional n-gram models. In speech recognition, class-based n-gram models have the potential to replace traditional models by addressing the challenges posed by large vocabulary sizes in real-world scenarios. By grouping words into classes, the models can generalize better and handle out-of-vocabulary words more effectively.

Exponential Language Models

Exponential MaxEnt language models [51] are a type of language model that use the principle of maximum entropy to estimate probabilities of generated sequences of words. The maximum entropy principle, or MaxEnt, states that in the absence of any other information, the model should distribute probabilities uniformly over all possible outcomes that are consistent with the observed data. This allows the model to assign probabilities in a way that avoids making unjustified assumptions.

Exponential MaxEnt models are particularly useful in ASR for language modeling, where they help improve the accuracy of transcription by predicting which words are more likely to occur given a particular sequence of words. Unlike n-gram based models, which rely on the assumption of local dependencies between words, exponential MaxEnt models capture long-range dependencies and can learn complex patterns in the data. They make use of a larger context window, which allows them to capture more global information and produce more accurate predictions for the next word.

According to the MaxEnt framework, the probability of a word w_i given the context history $h = w_{i-k}, w_{i-k+1}, \ldots, w_{i-1}$ is given by:

$$P(w_i|h) = \frac{\exp(\phi(h, w_i) \cdot \theta)}{Z(h, \theta)} \tag{4.22}$$

where θ is the parameter vector, $\phi(h, w_i)$ is the feature vector, and Z is the normalization constant over the vocabulary V:

$$Z(h, \theta) = \sum_{v \in V} \exp(\phi(h, v) \cdot \theta) \tag{4.23}$$

One of the main differences between exponential MaxEnt models and n-gram models is the way they handle feature extraction. Exponential models use a set of manually-defined features that capture different linguistic properties, such as word identity, part-of-speech, and context. These features are then combined linearly using weights that are learned during training. On the other hand, n-gram models use counts of word sequences to estimate probabilities, relying on the frequency of occurrence without explicitly incorporating linguistic features. Exponential Max-Ent models have the advantage of being able to generalize well to unseen contexts and handle data sparsity better than n-gram models. However, they require more computational resources and training data due to the larger context window and the need for feature engineering.

The training process for MaxEnt models involves iterating over the training corpus and updating the weights of the learned features to maximize the likelihood of the observed data. This can be achieved through various optimization algorithms, such as iterative scaling or stochastic gradient descent. The number of features, the size of the context window, and the complexity of the linguistic properties considered all contribute to the computational requirements. Additionally, the need for iterative optimization algorithms further adds to the training time.

Neural Language Models

Neural language models have emerged as a powerful approach in the field of automatic speech recognition for language modeling. Instead of relying on hand-crafted linguistic features, neural language models are designed to learn distributed representations of words using word embeddings. Word embeddings encode semantic properties of words in a continuous vector space, allowing the model to capture relationships and similarities between words. These similarities may be difficult to represent explicitly using manual feature engineering, and in case of word embeddings the similarities are trained directly from the data. Traditionally, word embeddings, such as word2vec [36], GloVe [45], or Pointcaré embeddings [39] represent the semantic similarity between words based on large corpora of written text, which is not necessarily robust in the presence of errors introduced by the acoustic model. An interesting idea has been proposed by Bengio and Heigold [2] to train the embeddings based on the similarity of word pronunciation. Similar ideas led to the development of alternative embedding methods, such as audio word2vec [10] or speech2vec [9].

Neural language models have a rich history of development, starting with the introduction of recurrent neural networks (RNNs) for language modeling [35]. RNNs were able to capture and model sequential dependencies within the input data by maintaining hidden states that pass information between time steps. However, RNNs suffered from the vanishing gradient problem, limiting their ability to handle long-range dependencies. To address this issue, LSTM networks were introduced [25]. LSTM networks incorporated memory cells and gating mechanisms, allowing them to learn and propagate information over longer sequences. This architectural improvement greatly enhanced the performance of neural language models. Another development came with the introduction of gated recurrent units (GRUs) [8], which simplified the LSTM architecture while still maintaining connectivity and mitigating the vanishing gradient problem. Another type of architecture commonly used is the Transformer model, which utilizes self-attention mechanisms [58] to capture contextual relationships across the input text.

Language modeling using neural networks is an extensive and complex field that is beyond the scope of this book, and we do not try to provide a comprehensive coverage of neural language modeling. However, it is worth highlighting some key features that make neural language models valuable in automatic speech recognition. One significant advantage of neural language models in ASR is their ability to improve the overall accuracy of transcription. By capturing rich contextual information and semantic relationships between words, these models can provide more accurate predictions for the next word. Due to their capacity, neural language models have the potential to correct mistakes made by the acoustic model. This capability is especially useful in scenarios where the acoustic model may misclassify certain words or encounter noise or distortion in the audio input. On the other hand, it is important to consider the trade-off between the accuracy of neural language models and the computational resources required to run the inference. Neural network-based models can be computationally demanding, especially with larger architectures and complex input representations, such as Transformer models. The computational intensity of running these models may necessitate significant computing power, limiting the feasibility of their deployment in certain resource-constrained environments.

An important limiting factor for neural language models in ASR is their inference time. ASR systems often require near real-time speed to facilitate interactive applications or live transcription. While neural language models have demonstrated impressive performance, their computational complexity can result in longer inference times. This can be a challenge for ASR applications that demand low latency. It is less likely that neural models will be used as standalone components in current ASR architectures. Instead, they are more likely to be integrated as part of an end-to-end solution [32] to speech recognition. Recent advances in deep learning, particularly in the field of sequence-to-sequence models, have paved the way for end-to-end ASR systems that can directly convert speech into text without the need for intermediate components such as language models or pronunciation dictionaries. These end-to-end architectures combine acoustic models, language models, and the decoding process into a single neural network, enabling a more streamlined and efficient approach to ASR. End-to-end speech recognition has shown promising results and offers the potential for more accurate transcription in various applications.

4.3.5 Decoding

The last step in the ASR pipeline is decoding, i.e., production of the final transcript on the basis of input raw audio signal converted to audio features. The task is to find

$$W^* = \arg\max_W P(X|W)P(W)I^{|W|} \tag{4.24}$$

where $I^{|W|}$ is the penalty for inserting a word hypothesis into the recognized transcription. Sometimes, the decoder may generate hypotheses that do not correspond to the actual speech, leading to incorrect transcriptions. These incorrect hypotheses often include the insertion of additional words that were not present in the original speech. To address this issue, a word insertion penalty is introduced to discourage the insertion of such unnecessary or incorrect words. The penalty is usually applied by reducing the scores of hypotheses with inserted words, making them less likely to be included in the final transcription. The value of the word insertion penalty can be adjusted based on the specific requirements and characteristics of the speech recognition system and the application domain.

The main problem though is the huge search space induced by the size of the vocabulary. The solution employed in commercial ASR systems is the *beam search* procedure, which maintains k decoding states (k is often referred to as the *width* of the beam) that represent possible hypotheses. As the audio input features are fed into ASR, the hypotheses with low probability are discarded to make room for more probable sequences of words. The algorithm proceeds as follows. The acoustic features are aggregated across frames until the end of a word is detected. The language model adds the score representing the probability of the word, and the word insertion penalty is applied. Word hypotheses are sorted based on their total scores and the beam search is pruned by eliminating low probability hypotheses. Apart from the first word in the hypothesis, the remaining words are scored using conditional probability of the language model, i.e., $P(w_i, w_{i-1}, \dots, w_1|W) = P(w_1|W)P(w_i, w_{i-1}, \dots, w_2|w_1, W)$. This allows to maintain the context of the utterance where preceding words influence the probability of recognizing following words in the utterance. Beam search is a simple and efficient exploration of the ASR search space, providing balance between computational complexity of the task, and the available resources.

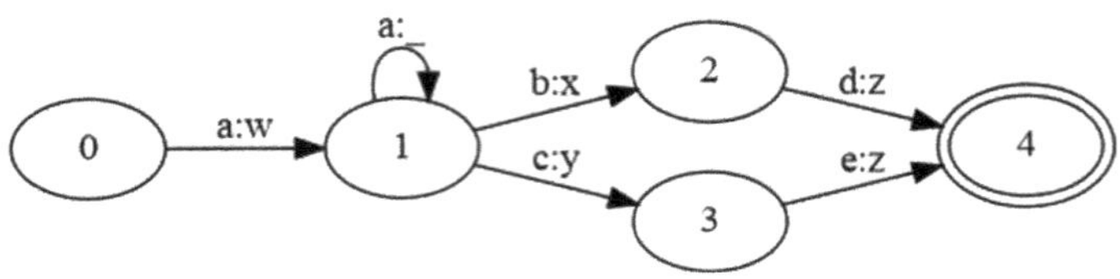

Fig. 4.4 Example of a finite state transducer

An alternative approach to transcript decoding is to use finite state transducers. Finite State Transducers (FSTs) are mathematical models that represent a set of input-output relationships between sequences of symbols. It is a type of automaton that consists of states, transitions, input symbols, output symbols, and weights associated with transitions (see Fig. 4.4 for an example of an FST). FST can be thought of as a machine that takes an input sequence and produces an output sequence based on a set of predefined rules. It can represent various types of mappings, such as the transformation of one sequence to another, string matching, or linguistic processes like transliteration or speech recognition. The states in an FST represent different stages or conditions during the processing of input sequence. Transitions connect these states, indicating the possible transformations from one state to another based on the input symbols. Each transition can also have an associated output symbol, which specifies the corresponding output generated during the transition. FSTs can be represented as directed graphs, where states are depicted as nodes, and transitions are depicted as labeled edges between the nodes. The weights assigned to each transition represent probabilities, costs, or other measures associated with that specific transformation. The key advantage of FSTs is their ability to efficiently encode and compactly represent large sets of mappings or rules. They can be composed, intersected, and combined with other FSTs to create more complex models.

In the context of ASR, FSTs can be used to model the mapping between acoustic features and word sequences. To produce transcripts using FSTs, the FST is constructed with the help of linguistic resources such as lexicons, language models, and acoustic models. Unlike beam search, which explores a limited number of hypotheses at each step, FST-based decoding computes the likelihood of all possible word sequences directly using the FST. This approach reduces the need for beam pruning, allowing for a more exhaustive search over the entire hypothesis space. The FST is designed to efficiently traverse all possible paths, taking advantage of its compact and deterministic representation. By incorporating weighted transitions at each state, FST-based decoding can assign scores to the potential word sequences, just like beam search. When compared with the beam search, FST-based decoding explicitly models linguistic and acoustic aspects of the ASR, allowing for easier debugging and interpretability. FSTs can handle larger and more complex language models, so the ASR can go beyond n-grams and use neural language models. Also, FST-based decoding can be optimized through techniques like determinization and minimization, resulting in faster and more memory-efficient computations.

FSTs can be used to express any model used in the ASR pipeline, which, given the fact that FSTs can be combined, allows to implement the entire pipeline using FST compositions. For instance, assume that three models have been trained. Model M_H accepts a sequence of Hidden Markov Model states (computed from audio features) and transforms it into phones. Model M_P implements a pronunciation lexicon by accepting a sequence of phones and outputing a sequence of words. Finally, model M_L is a language model implementing a word-level grammar. Then, by composing $M_H \circ M_P \circ M_L$ we can translate directly the input sequence of Hidden Markov Model states into the output transcript.

The result of the decoding is a set of transcript hypotheses, often conflicting with one another. The last step of the decoding phase is transcript rescoring, where based on confidence measures computed over the lattice of possible transcripts, the most probable final transcript is constructed. A *word lattice* is a graphical representation capturing multiple alternative paths between possible transcript hypotheses [40]. Consider the word lattice presented in Fig. 4.5 (symbol ϵ represents silence). Even though individual fragments of the output sequence are well captured by conditional probabilities of the language model ("*I think your*", "*uhm thank you*"), the choice of the final transcript is still hard. The utterance can be transcribed as "*uhm thank you for your patience*", but it might also get transcribed as "*I think your four your patents*". Thus, the lattice must be traversed to compute final confidence measures to select the best hypothesis. Due to possible explosion of the size of the word lattice, often the lattice is trimmed to contain only n-best hypotheses. To speed up computations, word lattices are compressed to *word confusion networks*, which are specialized word lattices [21, 34]. Word confusion networks capture the alternative hypotheses and explicitly represent confusion or uncertainty in the recognition process. Each node in the network represents a potential word hypothesis, including possible substitutions, insertions, deletions, and other types of errors (see Fig. 4.6). Word confusion networks can be utilized to perform lattice-based rescoring, incorporating knowledge about the possible confusion patterns. This approach improves the robustness and accuracy of the final transcripts by allowing to encode domain specific substitutions, such as "*eye phone*" → "*iphone*".

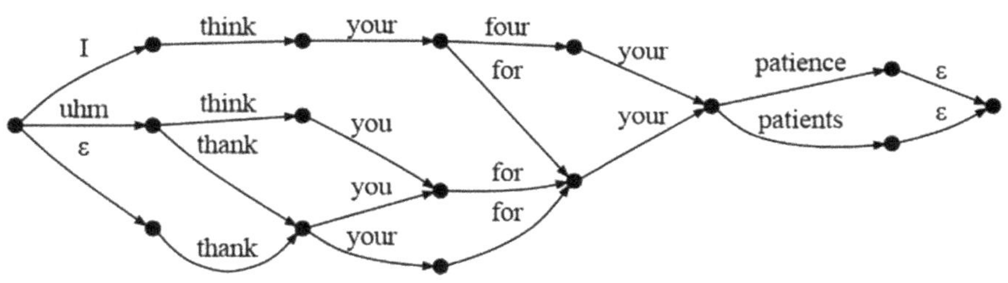

Fig. 4.5 Example of a word lattice

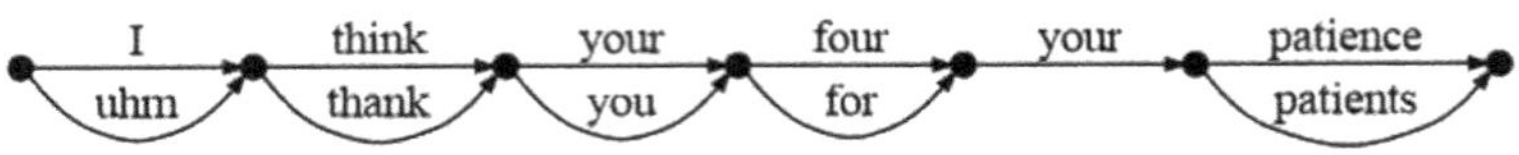

Fig. 4.6 Example of a word confusion network

Finally, we should mention an approach which assumes multiple passes over the hypothesis space. The multi-pass architecture of ASR transcript recognition follows a step-by-step process that combines different models to iteratively refine and improve the accuracy of the transcription. Initially, a simple monophone acoustic model (which considers individual phonetic units) and a basic language model are used to generate an initial word lattice. In the next step, the word lattice is rescored using a more advanced triphone acoustic model (which considers context-dependent phonetic units) and a more complex language model. This rescoring process involves reevaluating and assigning new scores to the hypotheses present in the lattice. The resulting set of hypotheses is then used to generate an n-best hypotheses list, capturing the most likely transcription alternatives. This n-best hypotheses list is further processed using domain-adapted acoustic and language models, which are specifically trained to handle the characteristics and specificities of the target domain or application. The domain-adapted models help refine the hypotheses by incorporating domain-specific knowledge and further reevaluating their likelihoods. The n-best hypotheses list is then rescored and reranked based on the scores provided by the domain-adapted models. The final outcome is a refined list of hypotheses, ordered based on their likelihoods.

4.4 Evaluation Metrics for ASR Systems

Evaluation metrics are used for assessing the performance and accuracy of ASR systems. However, evaluating the quality of transcriptions for spontaneous speech poses some unique challenges. One major challenge is the lack of sufficiently annotated validation sets for spontaneous speech. Obtaining accurate reference transcriptions for evaluation requires expensive manual annotation and is very time-consuming. The annotated transcript is often internally inconsistent (due to the complexity of the annotation task), so one has to carefully decide if a difference in reported evaluation metric reflects a true difference in the quality of compared ASRs, or if it is an artefact of the particular annotation. Another important aspect to consider in evaluating ASR systems for spontaneous speech is the impact of recognition errors on the overall quality of the transcript. Recognition errors can occur at different points in the transcription, and their impact can vary significantly depending on where the error has occurred. For example, errors in highly important content words (e.g., verbs) or critical information (e.g., named entities) can have a substantial impact on the overall understanding of the transcript. On the other hand, errors in non-essential words (e.g., prepositions) or less important phrases (e.g., back-channeling) may have a lesser impact on the transcript's comprehension. Below we present a brief overview of popular metrics used to evaluate the quality of ASR systems.

Word Error Rate

Word Error Rate (WER) is the primary evaluation metric used in ASR to measure the accuracy of transcripts. WER provides a quantitative measure of the level of agreement between the recognized output from an ASR system and the reference transcript. The main idea behind WER is to calculate the number of word-level substitutions, insertions, and deletions needed to transform the recognized transcript into the reference transcript. This metric allows to compare different ASR systems or assess the performance of a single system under different conditions.

Using WER as an evaluation metric does pose several problems. The most consequential is the problem of proper word alignment, mainly when dealing with spontaneous speech or unclear boundaries between words. ASR systems can sometimes misalign words, resulting in errors that affect the WER. In cases where word boundaries are ambiguous or misaligned, accurately assessing the error rate becomes difficult. Another problem arises when there is an overlap of speech from multiple parties in a conversation. Distinguishing and segmenting speech from different speakers is not a trivial task, and ASR systems may struggle to accurately transcribe overlapping speech. This can lead to inflated error rates in the WER calculation.

Despite these problems, WER remains the most widely used evaluation metric for ASR. It provides a metric that is easy to understand and compare, facilitating performance assessments and enabling progress measurements in the field of ASR. A benchmark of available ASRs [56] cites word error rates as low as 2%–3% on standard datasets. This may create a false conviction that speech recognition is mostly a solved problem. In reality, these reports are grossly over-estimated, as pointed out by Szymanski *et al.* [57]. The authors claim that the reason for this misconception is the confounding of two regimes of speech recognition. The majority of interactions with ASRs happen in the context of chatbot-like interactions, when a human is fully aware of speaking to a machine. In these circumstances most people simplify their utterances, speaking in short, well-structured phrases which obey the correct grammatical structure of either an interrogative or an imperative sentence. This is in stark opposition to spontaneous human conversations. As a matter of fact, these two regimes are different enough to build an 80% accuracy classifier to distinguish between them based solely on acoustic features [54]. The authors perform a real-world benchmark and obtain strikingly different results from the published benchmarks conducted on publicly available datasets, such as LibriSpeech [41], WSJ [43], CallHome, Fisher [11], and Switchboard [18]. The subject of evaluation involved several ASR systems on a multi-domain dataset derived from 50 call center conversations. These conversations were recorded at a standard, modern telephony quality of 8 kHz. The dataset encompassed 8.5 hours of audio, which incorporated 2.2 hours of speech. It was composed of 1595 agent and 1361 customer utterances, with an average utterance containing approximately 10 words, resembling a sentence-length

Table 4.1 Real-world WER benchmark [%] (source: [57])

	ASR 1	ASR 2	ASR 3
Booking	21.19	22.16	20.95
Finance	16.82	18.46	15.83
Insurance 1	18.01	20.20	17.84
Insurance 2	15.25	17.11	13.73
Telecomm.	19.75	23.31	17.62
Agent	16.97	17.83	16.49
Customer	17.87	20.99	16.48

phrase. The method of speech recognition employed relied on three differing state-of-the-art commercial ASR solutions evaluated as representative of what is largely accessible in the market (for legal reasons the authors could not disclose their names).

The gap between the state-of-the-art results reported on public benchmarks and real-life data is remarkable. On average, one in every five words in the output transcription is wrong. Table 4.1 presents WERs obtained on the real-world benchmark. The five domains evaluated in the benchmark include booking touristic reservations, finance, two types of insurance domains, and telecommunication conversations. One can see that insurance #2 was the easiest of domains for the ASRs, followed by finance. The highest error rates are obtained for booking and wireless telecommunication calls, which can be related to the fact that more of their utterances contain entities related to date and time, money, places, and product or company names. The bottom part of the table presents the division of WERs per channel. These differences are not large enough to assume that one of the channels is in a significantly better position for performing downstream NLP tasks.

Results presented in Table 4.1 clearly indicate the shortcomings of current benchmarks of ASR systems. New audio datasets have to be collected and annotated, providing better alignment of benchmarks with current application domains of ASR systems. The benchmarks should also challenge the development of more inclusive acoustic models representing a much broader spectrum of dialects, account for technological advances which influence physical properties of processed audio signals, and develop language models for multi-domain conversations. As the authors note, it is hard to believe, but most available spontaneous conversation datasets are 20 years old, and their quality of annotation is questionable. Often, part-of-speech tags and dependency structure are missing, and dialog act annotation is rare. Annotations of any classes of named entities (beyond the most rudimentary labeling scheme of `LOC`, `ORG`, `PER`, and `MISC`) are practically non-existent. For instance, the LDC catalog, which is the most comprehensive repository of language-related corpora, lists only four datasets of spontaneous conversations that are not derivative of Fisher, CallHome, or SwitchBoard datasets: LDC2015S08, LDC2019S06, LDC2013S05, and LDC2010S02. Unfortunately, these datasets miss rich annotations and they do not represent real-world, multi-domain conversations. The authors also mention recent attempts to improve the quality of ASR benchmarks by artificial generation of spontaneous speech datasets via text-to-speech [30, 31], but this approach solves the problem only to a certain degree. Synthetically generated audio streams lack many subtle characteristics of human speech, such as prosodic features and non-verbal acoustic cues. Text-to-speech synthesizers do not reproduce many aspects of human speech, and their output is not well aligned with real-world audio streams to which ASR systems will be applied.

Keyword Error Rate

Keyword Error Rate (KER) is an alternative metric used to evaluate the accuracy and performance of ASR systems. Unlike WER, which focuses on the overall accuracy of the entire transcription, KER specifically measures the accuracy of recognizing and transcribing predetermined keywords or phrases [37, 42]. This targeted evaluation allows for a more precise assessment of ASR systems, especially in applications where specific terms or commands are important. The advantage of KER is its ability to emphasize the recognition and transcription of keywords, making it particularly useful for keyword-driven applications like voice assistants. However, KER's main disadvantage is its limited scope, as it overlooks errors in non-keyword regions of the transcription. On the other hand, WER provides a complete evaluation of the entire transcription but may not give specific emphasis to critical keywords. A combined evaluation using both KER and WER provides a more nuanced understanding of the ASR system's quality and performance.

Named Entity Error Rate

Named Entity Error Rate (NERWER) is a metric designed to evaluate the accuracy of ASR systems in recognizing and transcribing named entities. NERWER extends the basic evaluation of WER to include a broader range of named entities such as person names, locations, organizations, and dates [5]. Compared to WER, NERWER offers a more focused evaluation that specifically measures the recognition and transcription accuracy of named entities, which are often important in practical applications. This metric acknowledges the importance of accurately capturing named entities, allowing for more precise evaluation in scenarios where the correct extraction of these entities is required, such as in transcribing medical reports, legal proceedings, or financial transactions. Unlike KER, which assesses only specific keywords, NERWER provides a broader evaluation by considering a wider range of named entities. Additionally, evaluation using KER requires manual annotation of keywords in golden transcripts (ground truth transcripts). For NERWER, dedicated NER models can be applied to automatically tag named entities in the transcript, significantly decreasing the need for manual human annotation. However, NERWER has limitations. It may not capture errors in non-named entity regions, potentially providing an incomplete evaluation of overall ASR system performance. The performance of entity recognition algorithms may affect the accuracy of NERWER. Hence, the limitations of the underlying NER algorithm should be taken into account when interpreting NERWER scores.

Phonetically Oriented Word Error Rate

Phonetically Oriented Word Error Rate (POWER) measures the percentage of incorrectly recognized phonemes in a transcription [52]. By measuring the errors in recognizing and representing phonemes, POWER provides insights into the system's ability to correctly interpret and transcribe spoken language. The main advantage of POWER over WER is that it focuses on the acoustic challenges of speech recognition. By evaluating phonemic accuracy, POWER captures errors caused by phonetic variations, dialects, accents, and coarticulation effects that may not be reflected by WER alone. POWER is especially useful in tasks where phonetic details are important, such as language learning, speech therapy, or phonetic research. A scenario where the use of POWER is warranted is in speech therapy applications. Precise recognition and transcription of phonemes are essential for evaluating a person's speech production and progress. By using POWER, therapists can assess the ASR system's accuracy in capturing specific phonetic details, allowing them to tailor interventions and track improvements in an individual's pronunciation and articulation. POWER provides a more targeted evaluation in this context, aiding in the development and refinement of speech therapy tools and techniques. However, a disadvantage of POWER is its complexity compared to WER. Calculating POWER requires mapping the recognized phonemes to reference phonemes, which results in additional linguistic information processing. This complexity can make the evaluation process more involved and necessitate specialized knowledge and resources.

Speaker Error Rate

Speaker Error Rate (SER) measures the percentage of incorrectly recognized speakers in a transcript. SER takes into account errors made in determining the correct speaker identity in a multi-speaker audio stream or dialogue. SER is not an alternative metric to WER as it does not reflect the quality of the transcript, it only assesses the ASR system's ability to correctly recognize and differentiate between speakers. This metric captures errors that arise when multiple speakers are present, such as misattributing speech segments or inaccurately distinguishing between speakers with similar speech characteristics. Practical application domain of SER are meeting transcriptions, where understanding who said what is needed for accurate minutes or analysis. Similarly, in call center analytics, SER helps measure the system's accuracy in recognizing different speakers during customer interactions.

Levenshtein Distance

Levenshtein distance is a metric used to quantify the difference or similarity between two strings. It measures the minimum number of single-character edits required (insertions, deletions, or substitutions) to transform one string into another. The basic idea behind Levenshtein distance is to calculate the number of operations needed to make the strings identical. In the context of ASR evaluation, Levenshtein distance can be used to compare the recognized transcript generated by an ASR system with the reference transcript, allowing for the calculation of the error rate [16, 59]. One advantage of using Levenshtein distance is its ability to pinpoint specific errors and highlight the accuracy of the ASR system at the character level. This can lead to a more comprehensive understanding of the system's performance, particularly in scenarios where phonetic variations or mispronunciations are important factors. However, the disadvantage of the Levenshtein distance is that it can be more computationally intensive compared to WER since it operates at the character level, potentially increasing the evaluation time. Additionally, the interpretation of Levenshtein distance results requires careful consideration, as some edits may have varying degrees of impact on the overall transcription quality. Scenarios where using Levenshtein distance might be preferred over WER include applications where character accuracy is important, such as transcribing text messages or short utterances. It can also be useful for analyzing the performance of ASR systems with specific challenges, such as strong accents, non-standard pronunciations, or noisy environments.

4.5 Closing

The future of ASR shows several promising research directions. One such direction is streaming ASR, which aims to provide instantaneous transcription by processing speech in a continuous stream. Streaming ASR seeks to eliminate any significant delay between spoken input and transcription output, making it ideal for applications requiring real-time interaction. The advancement of faster processing capabilities and efficient algorithms will play an important role in enabling real-time ASR, leading to applications like live captioning, voice assistants, and telephony services. There is also a growing interest in deploying ASR on edge devices, such as smartphones or smart speakers, which brings the benefits of low latency, privacy preservation, and reduced dependence on cloud resources. The development of end-to-end ASR systems, involving integrated acoustic and language models without intermediate components, opens the way for more accurate transcriptions. Future research in ASR will undoubtedly focus on these areas to further enhance the speed, accuracy, and adaptability of speech recognition systems in various domains.

The development of robust ASR systems faces several challenges. One significant challenge is training ASR systems for different languages and dialects. Each language and dialect comes with its own unique characteristics, vocabulary, and phonetic variations, and training ASR requires the collection of a large annotated dataset. For low-resourced languages and dialects this can be prohibitively expensive. Another challenge is using ASR in noisy environments. Background noise, reverberation, and other acoustic factors can degrade the quality of the audio input, affecting the accuracy of transcription. Techniques that can handle noise and enhance speech signals are essential for improving ASR performance in real-world settings.

As with any technology, ASR raises ethical concerns, particularly when it comes to its use in surveillance and wiretapping. The ability of ASR systems to process and interpret spoken content can potentially infringe upon privacy rights. As ASR continues to advance, it is necessary to ensure robust security measures, lawful use, and transparent guidelines to protect user privacy and prevent misuse of the technology.

References

1. N. Ahmed, T. Natarajan, K.R. Rao, Discrete cosine transform. IEEE Trans. Comput. **100**(1), 90–93 (1974)
2. S. Bengio, G. Heigold, Word embeddings for speech recognition (2014)
3. J. Bilmes, K. Kirchhoff, Factored language models and generalized parallel backoff, in *Companion Volume of the Proceedings of HLT-NAACL 2003-Short Papers* (2003), pp. 4–6
4. P.F. Brown, V.J. Della Pietra, P.V. Desouza, J.C. Lai, R.L. Mercer, Class-based n-gram models of natural language. Comput. Linguist. **18**(4), 467–480 (1992)
5. J.D. Burger, D.D. Palmer, L. Hirschman, Named entity scoring for speech input, in *36th Annual Meeting of the Association for Computational Linguistics and 17th International Conference on Computational Linguistics, Volume 1* (1998), pp. 201–205
6. W. Chan, N. Jaitly, Q. Le, O. Vinyals, Listen, attend and spell: A neural network for large vocabulary conversational speech recognition, in *2016 IEEE international conference on acoustics, speech and signal processing (ICASSP)* (IEEE, 2016), pp. 4960–4964
7. C. Chelba, F. Jelinek, Structured language modeling. Comput. Speech Lang. **14**(4), 283–332 (2000)
8. J. Chung, C. Gulcehre, K. Cho, Y. Bengio, Empirical evaluation of gated recurrent neural networks on sequence modeling. *arXiv preprint* arXiv:1412.3555 (2014)
9. Y.-A. Chung, J. Glass, Speech2vec: A sequence-to-sequence framework for learning word embeddings from speech. *arXiv preprint* arXiv:1803.08976 (2018)
10. Y.-A. Chung, C.-C. Wu, C.-H. Shen, H.-Y. Lee, L.-S. Lee, Audio word2vec: Unsupervised learning of audio segment representations using sequence-to-sequence autoencoder. *arXiv preprint* arXiv:1603.00982 (2016)
11. C. Cieri, D. Miller, K. Walker, The Fisher corpus: a resource for the next generations of speech-to-text. Int. Conf. Lang. Resour. Eval. **4**, 69–71 (2004)
12. R. Collobert, C. Puhrsch, G. Synnaeve, Wav2letter: an end-to-end convnet-based speech recognition system. *arXiv preprint* arXiv:1609.03193 (2016)
13. A. Deri, K. Knight, Grapheme-to-phoneme models for (almost) any language, in *Proceedings of the 54th Annual Meeting of the Association for Computational Linguistics (Volume 1: Long Papers)* (2016), pp. 399–408
14. N. Desai, K. Dhameliya, V. Desai, Feature extraction and classification techniques for speech recognition: A review. Int. J. Emer. Technol. Adv. Eng. **3**(12), 367–371 (2013)
15. G. Fant, *Acoustic Theory of Speech Production. Description and Analysis of Contemporary Standard Russian* (Mouton, 1960)
16. J.G. Fiscus, J. Ajot, N. Radde, C. Laprun, et al., Multiple dimension levenshtein edit distance calculations for evaluating automatic speech recognition systems during simultaneous speech, in *LREC* (Citeseer, 2006), pp. 803–808

17. D. Gildea, T. Hofmann, Topic-based language models using em, in *Sixth European Conference on Speech Communication and Technology* (1999)
18. J.J. Godfrey, E.C. Holliman, J. McDaniel, Switchboard: Telephone speech corpus for research and development, in *[Proceedings] ICASSP-92: 1992 IEEE International Conference on Acoustics, Speech, and Signal Processing*, volume 1 (IEEE, 1992), pp. 517–520
19. A. Graves, A. Graves, Connectionist temporal classification, in *Supervised sequence labelling with recurrent neural networks* (2012), pp. 61–93
20. A. Graves, N. Jaitly, Towards end-to-end speech recognition with recurrent neural networks, in *International conference on machine learning* (PMLR, 2014), pp. 1764–1772
21. D. Hakkani-Tür, F. Béchet, G. Riccardi, G. Tur, Beyond asr 1-best: Using word confusion networks in spoken language understanding. Comput. Speech Lang. **20**(4), 495–514 (2006)
22. A. Hannun, C. Case, J. Casper, B. Catanzaro, G. Diamos, E. Elsen, R. Prenger, S. Satheesh, S. Sengupta, A. Coates, et al., Deep speech: Scaling up end-to-end speech recognition. *arXiv preprint* arXiv:1412.5567 (2014)
23. R. Hibare, A. Vibhute, Feature extraction techniques in speech processing: a survey. Int. J. Comput. Appl. **107**(5) (2014)
24. G. Hinton, L. Deng, D. Yu, G.E. Dahl, A.-R. Mohamed, N. Jaitly, A. Senior, V. Vanhoucke, P. Nguyen, T.N. Sainath et al., Deep neural networks for acoustic modeling in speech recognition: The shared views of four research groups. IEEE Sig. Process. Mag. **29**(6), 82–97 (2012)
25. S. Hochreiter, J. Schmidhuber, Long short-term memory. Neural Comput. **9**(8), 1735–1780 (1997)
26. B.H. Juang, L.R. Rabiner, Hidden markov models for speech recognition. Technometrics **33**(3), 251–272 (1991)
27. B.-H. Juang, L.R. Rabiner, Automatic speech recognition–a brief history of the technology development. *Georgia Institute of Technology. Atlanta Rutgers University and the University of California. Santa Barbara*, 1:67 (2005)
28. S. Kanthak, H. Ney, Multilingual acoustic modeling using graphemes, in *INTERSPEECH* (2003)
29. R. Lau, R. Rosenfeld, S. Roukos, Trigger-based language models: A maximum entropy approach, in *1993 IEEE International Conference on Acoustics, Speech, and Signal Processing*, volume 2 (IEEE, 1993), pp. 45–48
30. C.-H. Lee, S.-M. Wang, H.-C. Chang, H.-Y. Lee, Odsqa: Open-domain spoken question answering dataset, in *2018 IEEE Spoken Language Technology Workshop (SLT)* (IEEE, 2018), pp. 949–956
31. C.-H. Li, S.-L. Wu, C.-L. Liu, H.-y. Lee, Spoken squad: A study of mitigating the impact of speech recognition errors on listening comprehension. *arXiv preprint* arXiv:1804.00320 (2018)
32. J. Li et al., Recent advances in end-to-end automatic speech recognition. APSIPA Trans. Sig. Inf. Process. **11**(1) (2022)
33. H. Liu, S. Jin, C. Zhang, Connectionist temporal classification with maximum entropy regularization. Adv. Neural Inf. Process. Syst. **31** (2018)
34. L. Mangu, E. Brill, A. Stolcke, *Finding consensus among words: lattice-based word error minimization* (In Eurospeech, Citeseer, 1999)
35. T. Mikolov, M. Karafiát, L. Burget, J. Cernockỳ, S. Khudanpur, Recurrent neural network based language model, in *Interspeech*, volume 2 (Makuhari, 2010), pp. 1045–1048
36. T. Mikolov, I. Sutskever, K. Chen, G.S. Corrado, J. Dean, Distributed representations of words and phrases and their compositionality. Adv. Neural Inf. Process. Syst. **26** (2013)
37. H. Nanjo, T. Kawahara, A new asr evaluation measure and minimum bayes-risk decoding for open-domain speech understanding, in *Proceedings.(ICASSP'05). IEEE International Conference on Acoustics, Speech, and Signal Processing, 2005.*, volume 1 (IEEE, 2005), pp. I–1053
38. A.B. Nassif, I. Shahin, I. Attili, M. Azzeh, K. Shaalan, Speech recognition using deep neural networks: A systematic review. IEEE Access **7**, 19143–19165 (2019)
39. M. Nickel, D. Kiela, Poincaré embeddings for learning hierarchical representations. Adv. Neural Inf. Process. Syst. **30** (2017)

40. M. Oerder, H. Ney, Word graphs: An efficient interface between continuous-speech recognition and language understanding, in *1993 IEEE International Conference on Acoustics, Speech, and Signal Processing*, volume 2 (IEEE, 1993), pp. 119–122
41. V. Panayotov, G. Chen, D. Povey, S. Khudanpur, Librispeech: an asr corpus based on public domain audio books, in *2015 IEEE International Conference on Acoustics, Speech and Signal Processing (ICASSP)* (IEEE, 2015), pp. 5206–5210
42. Y. Park, S. Patwardhan, K. Visweswariah, S.C. Gates, An empirical analysis of word error rate and keyword error rate. Interspeech **2008**, 2070–2073 (2008)
43. D.B. Paul, J.M. Baker, The design for the wall street journal-based csr corpus, in *Proceedings of the workshop on Speech and Natural Language* (Association for Computational Linguistics, 1992), pp. 357–362
44. A. Pauls, D. Klein, Faster and smaller n-gram language models, in *Proceedings of the 49th annual meeting of the Association for Computational Linguistics: Human Language Technologies* (2011), pp. 258–267
45. J. Pennington, R. Socher, C. Manning, GloVe: Global vectors for word representation, in *Conference on Empirical Methods in Natural Language Processing* (2014), pp. 1532–1543
46. C. Krishna, *Puvvada, Piotr Żelasko, He Huang, Oleksii Hrinchuk, Nithin Rao Koluguri, Kunal Dhawan, Somshubra Majumdar, Elena Rastorgueva, Zhehuai Chen, Vitaly Lavrukhin, Jagadeesh Balam, and Boris Ginsburg* (Accurate speech recognition & translation without web-scale data, Less is more, 2024)
47. A. Radford, J.W. Kim, T. Xu, G. Brockman, C. McLeavey, I. Sutskever, Robust speech recognition via large-scale weak supervision, in *International Conference on Machine Learning* (PMLR, 2023), pp. 28492–28518
48. K. Rao, F. Peng, H. Sak, F. Beaufays, Grapheme-to-phoneme conversion using long short-term memory recurrent neural networks, in *2015 IEEE International Conference on Acoustics, Speech and Signal Processing (ICASSP)* (IEEE, 2015), pp. 4225–4229
49. K. Rao, H. Sak, Multi-accent speech recognition with hierarchical grapheme based models, in *2017 IEEE international conference on acoustics, speech and signal processing (ICASSP)* (IEEE, 2017), pp. 4815–4819
50. D. Rekesh, N.R. Koluguri, S. Kriman, S. Majumdar, V. Noroozi, H. Huang, O. Hrinchuk, K. Puvvada, A. Kumar, J. Balam, et al., Fast conformer with linearly scalable attention for efficient speech recognition, in *2023 IEEE Automatic Speech Recognition and Understanding Workshop (ASRU)* (IEEE, 2023), pp. 1–8
51. R. Rosenfeld et al., A maximum entropy approach to adaptive statistical language modelling. Comput. Speech Lang. **10**(3), 187 (1996)
52. N. Ruiz, M. Federico, Phonetically-oriented word error alignment for speech recognition error analysis in speech translation, in *2015 IEEE Workshop on Automatic Speech Recognition and Understanding (ASRU)* (IEEE, 2015), pp. 296–302
53. P. Senin, Dynamic time warping algorithm review. Information and Computer Science Department University of Hawaii at Manoa Honolulu, USA **855**(1–23), 40 (2008)
54. I. Siegert, J. Krüger, How do we speak with alexa: Subjective and objective assessments of changes in speaking style between hc and hh conversations. Kognitive Systeme **1**, 2018 (2018)
55. K.N. Stevens, Toward a model for speech recognition. J. Acoustical Soc. America **32**(1), 47–55 (1960)
56. G. Synnaeve, Wer are we? https://github.com/syhw/wer_are_we (2020). [Online; accessed 6-December-2023]
57. P. Szymański, P. Żelasko, M. Morzy, A. Szymczak, M. Żyła-Hoppe, J. Banaszczak, L. Augustyniak, J. Mizgajski, Y. Carmiel, Wer we are and wer we think we are. *arXiv preprint* arXiv:2010.03432 (2020)
58. A. Vaswani, N. Shazeer, N. Parmar, J. Uszkoreit, L. Jones, A.N. Gomez, Ł. Kaiser, I. Polosukhin, Attention is all you need. Adv. Neural Inf. Process. Syst. **30** (2017)
59. B. Ziółko, J. Gałka, T. Jadczyk, D. Skurzok, Modified weighted levenshtein distance in automatic speech recognition, in *Proceedings of the 16th Conference on Applications of Mathematics in Biology and Medicine* (2010), pp. 116–120

Chapter 5
Transcript Engineering

5.1 Transcript Alignment

Token alignment in ASR transcripts refers to the task of mapping each word or phoneme in the transcribed text to its exact starting and ending points in the corresponding audio. This involves precisely aligning the transcribed tokens with the time-stamped audio file, ensuring that the written transcription matches the spoken dialogue both in content and sequence. The goal is to create a transcript that accurately reflects the temporal dynamics of the speech, including pacing, pauses, and inflections. This task is important for various applications like subtitle generation, speech analytics, and detailed linguistic studies. Token alignment in the production of speech transcripts presents a significant challenge, primarily due to the discrepancies between spoken language and its written representation. Natural speech often includes hesitations and repetitions that are not typically found in written text, making it difficult to align the two forms accurately. Two factors in particular make token alignment difficult: the overlap between speaking parties introduces a lot of parallelism between tokens, and the presence of non-speech may cause confusion. Consider the following example of a spontaneous conversation between two individuals (Fig. 5.1).

In this example, the overlap of speech, interruptions, and a quick back-and-forth between the speakers make it difficult to accurately align tokens in a transcript. Transcribing such conversations requires careful consideration of when each speaker begins and ends their utterance, and how the interruptions and overlaps are handled.

Most approaches use a version of the forced alignment algorithm. The method works on the principle of matching the phonetic transcription of a sentence with the corresponding speech signal, using acoustic models. The algorithm requires a lexicon or pronunciation dictionary, which provides the possible phonetic transcriptions of the words in the transcript. Each word is represented as a sequence of phonemes which, when pronounced, should align with the speech signal. The acoustic models used in forced alignment are trained to recognize these phonemes in the audio signal.

M. Morzy, *Spoken Language Processing*, Studies in Computational Intelligence 1205,
https://doi.org/10.1007/978-3-031-88566-2_5

Fig. 5.1 Spontaneous conversation with hesitations and overlaps

SPEAKER 1: so i uh i went to that um that new
SPEAKER 2: that um new place
SPEAKER 1: yeah place
SPEAKER 2: downtown right
SPEAKER 1: yeah yeah the new place and i uh i got um
SPEAKER 2: probably ramen right
SPEAKER 1: ra yeah ramen um it was really really good

The alignment process involves matching each phoneme in the phonetic transcription to a segment of the audio signal. The algorithm uses a search process to find the best match between the phonemes and the audio signal, considering both the similarity of the audio to the acoustic models of the phonemes and the timing of the phonemes in the sentence. The result of the forced alignment process is a time-aligned transcript, where each word or phoneme in the transcript is associated with a start and end time in the audio signal.

An example of the forced alignment is the recursive method introduced by Moreno *et al.* [29]. The key idea of this technique is to transform forced alignment into a series of speech recognition problems. During recursive alignment the dictionary and language model used in each recognition pass are progressively narrowed down. This helps the algorithm to focus on the most relevant words and phrases for each audio segment. The process starts by representing the audio file with features, such as mel-cepstrum vectors. Then, a dictionary and a language model are constructed from the text transcript. These elements are used to run a large vocabulary speech recognition system on the audio stream, generating a hypothesized text string. Next, the hypothesized text is aligned with the correct transcript using dynamic programming. This alignment is then analyzed to select *anchors* or segments with high confidence levels. These anchors are used to divide the audio and text segments into aligned and unaligned portions. The algorithm then proceeds recursively on each unaligned segment. For each segment, the dictionary and language model are narrowed to those words and phrases that appear in the segment's transcript. This focused approach improves alignment accuracy, especially in noisy conditions. The recursion continues until all segments are aligned or the termination condition is met. This condition could be a fully aligned text or a maximum duration for an unaligned segment. The recursive forced alignment technique offers several advantages. First, it concentrates on aligning confident segments first, postponing decisions on less certain portions until later iterations. This approach minimizes the impact of potential errors or uncertainties. Second, working with smaller segments in later iterations allows for more targeted language models. This improves alignment accuracy, particularly in noisy environments. Finally, the recursive method can handle errors or gaps in the transcript. By confining the impact of errors to smaller regions, the algorithm minimizes their overall effect on the alignment. In essence, the recursive forced alignment technique uses a divide-and-conquer approach to align audio segments with their transcripts. It progressively identifies and aligns confident segments, gradually reducing the scope of the search space and improving accuracy along the way.

An interesting variation is presented by Hazen [19] who introduces the pseudo-forced alignment procedure to correct human-generated transcripts of speech. Pseudo-forced alignment rectifies manual transcripts by combining ASR, alignment, and editing techniques. Firstly, ASR is applied with a transcript-guided language model, generating anchor points through transcript matching. Next, pseudo-forced alignment of transcripts is executed, enabling insertions, deletions, and substitutions. Finally, the alignment is edited by rerunning ASR on segments with potential errors, allowing word sequence proposals under language model constraints. This procedure enhances accuracy by discovering and correcting transcript errors during alignment. A similar method based on phoneme alignment is developed by Haubold and Kender [18]. Their alignment method aligns highly imperfect text transcriptions to their original speech signal by extracting phonemes from both the speech and text and aligning them using dynamic programming with an edit-distance algorithm. The cost of incurring copies and deletions is the same, while replacements and insertions are assigned a higher cost. Once the alignment is completed, time codes from individual phoneme matches are assigned to their original words, producing the temporal text-to-speech alignment. This method is shown to be effective for aligning highly imperfect transcripts with word error rates over 70%, with more than 60%, 75%, and 90% of all words aligned correctly within 10, 20, and 30-second error margins, respectively.

5.2 Transcript Annotation

An essential concept in transcript annotation is the notion of a dialog act, introduced in Sect. 3.3.1. Let us recall that a dialogue act refers to each individual utterance that serves a distinct function within the dialogue. Dialogue acts can take various forms, such as questions, remarks, statements, requests for reactions, and more. Categorizing portions of dialogue into specific dialogue acts is commonly referred to as *spoken utterance classification*. This process involves assigning predefined labels to different types of dialogue acts to facilitate analysis and interpretation. For instance, let us consider a conversation between two friends (Fig. 5.2).

In the conversation we observe several dialogue acts. The conversation begins with a greeting from John and a response from Sarah. John then sets the stage for a request for help by expressing a feeling of being lost. Sarah responds by offering assistance and asking for clarification on what John needs help with. John then presents an expression of doubt regarding his career change decision. Sarah follows up with a question to understand the reasons behind John's doubts. John further explains his doubts, which leads to Sarah asking for more information about his alternative career ideas. Finally, John seeks advice and guidance from Sarah, which she responds to by offering her support and suggesting further discussion.

There are numerous taxonomies of dialogue acts developed over the years for different domains of conversations. Several dialogue acts are shared across many taxonomies and can be found in most unscripted conversations. Apart from greeting,

JOHN: : Hey Sarah! Long time no see. How have you been?
SARAH: Hi John! I've been good, thanks for asking. How about you?
JOHN: : I've been alright, but I'm feeling a bit lost. Can you help me with something?
SARAH: Of course, what's going on?
JOHN: : Well, I've been thinking about changing my career, but I'm not sure if it's the right decision.
SARAH: Hmm, I understand. What's making you doubt your current path?
JOHN: : I feel like I'm not passionate about my job anymore, and I'm looking for something more fulfilling.
SARAH: I see where you're coming from. Have you thought about what alternative career path you might be interested in?
JOHN: : Yes, I've been considering starting my own business. Do you think it's too risky?
SARAH: It depends on your skills, market research, and financial stability. Have you done any planning or research so far?
JOHN: : Not much yet, but I'd appreciate any guidance or advice you can offer.
SARAH: Of course! Let's set up a time to meet and discuss your ideas further. Maybe we can come up with a solid plan together.

Fig. 5.2 Conversation with dialogue acts

request for help, request for clarification, expression of doubt, and question dialogue acts depicted above, other examples of dialogue acts include directive ("*Please pass me the salt.*"), disagreement ("*I don't think you are right*"), or offering assistance ("*If you need any help, just let me know*"). The range of dialogue acts can be very diverse in everyday conversations. Dialogue acts are slightly narrower than segments. *Segments* are topically cohesive and usually consist of several utterances, where both speakers agree on the subject and do not change the topic of conversation. In general, topical segmentation of transcripts is difficult and manual segmentation is usually highly subjective. There are some clues in prosodic features (e.g. silence and non-speech events may be longer between segments), but this can be deceiving. There is usually no notion of a single story that can be broken down into topics in a spontaneous conversation.

As we have mentioned, there are multiple taxonomies (or annotation schemes) of dialogue acts. Before presenting an example of a taxonomy, let us inspect the dimensions commonly used to define taxonomies.

1. Task/Activity: this dimension focuses on identifying dialogue acts related to the performance of specific tasks or activities within the conversation. Examples include requesting information ("*Can you tell me the time of the meeting?*") or giving instructions ("*Please make sure to submit the report by tomorrow.*")
2. Auto-feedback: this dimension involves providing feedback or information regarding the processing of the previous speaker's utterance. Examples include confirmation ("*Yes, I understand what you're saying.*") or interruption ("*Wait, let me clarify something.*")
3. Allo-feedback: this dimension encompasses dialogue acts where the speaker expresses opinions or solicits information about the addressee's processing of previous utterances. Examples include seeking clarification ("*Could you explain*

that in more detail?") and seeking agreement ("*Do you agree with what I just said?*")

4. Contact Management: this dimension focuses on maintaining contact and interaction during the conversation. Examples include checking if the other person is listening ("*Are you following?*") and establishing turn-taking norms ("*It's your turn to speak now.*")
5. Turn Management: this dimension involves acts related to the management of speaker roles, such as grabbing, keeping, yielding, or accepting the speaker role. Examples include initiating a turn ("*I'd like to share my thoughts on this.*") and passing the turn to someone else ("*Why don't you go ahead?*")
6. Time Management: this dimension addresses situations where the speaker requires more time to formulate their thoughts. Examples include using fillers to gain time ("*Um, let me think about it for a moment.*") or requesting a pause ("*Just a second*")
7. Discourse Structuring: this dimension involves acts related to the organization and structuring of the dialogue. Examples include opening the conversation ("*Hi, how can I help you today?*") and closing the conversation ("*It was nice talking to you. Goodbye!*")
8. Own Communication Management: this dimension focuses on acts where the speaker edits or modifies their own utterance. Examples include self-correction ("*I meant to say 'blue' not 'green'.*") or repeating for emphasis ("*I really, really hate Brussels sprouts.*")
9. Partner Communication Management: this dimension involves acts where the speaker assists or corrects the current speaker. Examples include providing clarification ("*Just to be clear, you're referring to the red file, right?*") or offering an alternative suggestion ("*Instead of pizza, why don't we just grab a hot-dog?*")
10. Social Obligations Management: this dimension encompasses acts related to the fulfillment of social conventions, such as greetings, apologies, and expressions of gratitude. Examples include greeting ("*Good morning! How are you doing today?*"), apologizing ("*I'm sorry for the confusion earlier.*"), or expressing gratitude ("*Thank you so much for your help.*")

Let us now examine an example dialogue act annotation scheme. DAMSL (Dialog Act Markup in Several Layers) taxonomy is a widely-used framework for annotating dialogue acts in conversational transcripts [1]. DAMSL is designed to provide a multi-layered annotation scheme that captures the richness and complexity of dialogue acts. The DAMSL taxonomy consists of three layers: the primary act, the cue act, and the mood act. The primary act represents the intended function of the speaker's utterance, such as questioning, stating, agreeing, or requesting. This layer is sometimes divided into forward-looking and backward-looking acts and represents the communicative functions of the conversation. The cue act captures additional contextual information, such as the speaker's attitude or intention, while the mood act indicates the speaker's emotional state or stance. The DAMSL taxonomy enables researchers and annotators to finely categorize dialogue acts, allowing for a more nuanced analysis of conversational structure and dynamics. It has been widely adopted in various domains, including NLP, computational linguistics, and discourse analysis, as a standard for transcript annotation.

Over the years, several criticisms against DAMSL structure have been raised. For instance, Bunt [5] observes that several dimensions of DAMSL do not qualify as proper dimensions, the selection of tags is questionable (some tags are mutually exclusive), and existing annotation schemes built on top of DAMSL do not make clear distinction between notions of a dimension and a layer. The criticism stems from the observation, that, in addition to striving to accomplish the main task or activity driving the conversation, the speaker must also keep track of mutual attention and ensure that understanding is maintained. Moreover, they bear the responsibility of managing various elements of the interaction, such as coordinating the roles of sender and receiver (turn management), navigating time limitations (time management), and initiating or concluding discussion topics (topic management).

For a dimension to qualify as a proper dimension, the following conditions must be met:

- an aspect of participating in a dialogue can be addressed by the means of dialogue acts that have a communicative function specific for that purpose,
- an aspect of participating in a dialogue can be addressed independently of other aspects, which means that the dialogue act can have a communicative function in one dimension and another function in other dimensions.

Also, the annotation scheme should take into consideration that several types of language functions, including questions, statements, answers, and requests, are not exclusive to any specific dimension but can be employed across multiple dimensions. These functions serve as general-purpose linguistic tools and their usage within a particular dimension depends on the meaning and intention conveyed by the utterance.

Based on these observations, Bunt proposes the following set of dimensions for dialogue act annotation: task/activity (open meeting, close meeting, appoint, hire), auto feedback (perception negative, overall positive), allo feedback (interpretation negative, evaluation elicitation), turn management (keeping, grabbing, giving), time management (stalling, pausing), contact management, own communication management, partner communication management, topic management, dialogue structuring, and social obligations (greeting, apology, thanking).

Independently of this set of dimensions there are general purpose functions:

- information seeking functions: yes/no questions, WH-questions, multiple choice questions
- information providing functions: informs, agreements, disagreements, corrections, answers, confirmations
- commissive functions: committing to an action, such as offers, promises, acceptance of requests
- directive functions: speaker wants the addressee to consider an action, such as instructions, request, suggestions

Another taxonomy for dialogue act annotation is the Dynamic Interpretation Theory (DIT) [4]. DIT focuses on capturing the dynamic nature of conversations. This taxonomy incorporates concepts such as speech acts, intentions, grounding, and referential meaning to provide a complete representation of dialogue acts. By utilizing a multidimensional approach, DIT enables the annotation of context-dependent aspects of communication. It allows researchers to gain insights into the underlying motivations, intentions, and pragmatics of dialogue acts, aiding in the analysis of conversational dynamics and facilitating the development of dialogue systems and NLP applications. DIT++ taxonomy builds upon its predecessor, by introducing additional features to enhance the annotation process [6]. One of its key features is the consideration of multiple dimensions. DIT++ incorporates dimensions such as speaker intentions, speech acts, grounding, and interaction effects to provide a more detailed analysis of dialogue acts. This taxonomy focuses on the dynamic aspects of conversations, allowing for the depiction of the evolving nature of dialogue and the influences between speakers.

The DiAML (Dialogue Act Markup Language) taxonomy [7] is a framework used for annotating dialogue acts in transcripts. It provides a structured and standardized system for categorizing and labeling the various speech acts present in conversations. DiAML encompasses a wide range of dialogue act classes, including questions, requests, statements, suggestions, agreements, disagreements, and more. It also incorporates additional dimensions such as information status, action type, and discourse segment. The taxonomy provides a flexible and extensible structure that can be adapted to different domains and dialogue contexts. By utilizing the DiAML taxonomy, researchers can effectively annotate dialogue acts in transcripts, facilitating various applications such as dialogue system development, natural language understanding, and dialogue analysis.

5.3 Segmentation

An important task in processing and refining the speech-to-text conversion is segmentation of transcripts from raw ASR output. The raw ASR transcript typically lacks punctuation marks and does not include any explicit indications of sentence or utterance boundaries. Consequently, there is a need for the computation of these borders to break down the transcript into meaningful segments. This process requires identifying pauses, intonation patterns, and other linguistic cues to determine where sentences or utterances begin and end. However, it can be challenging as it relies on contextual understanding and accurate interpretation of the speech content. Accurately segmenting transcripts is important for various downstream applications, where the structure and coherence of sentences matter.

Sentence segmentation in raw ASR output is challenging because traditional sentences are typically constructed based on strict grammar rules and syntactic structures of a prescriptive language. Spontaneous speech, especially in conversational contexts, deviates from these formal structures and conventions, making it harder

SPEAKER 1: I am sick of this, where is my package?!
SPEAKER 2: Sir, can you please
SPEAKER 1: It was supposed to be
SPEAKER 2: Sir, just let me
SPEAKER 1: like arriving yesterday and um
SPEAKER 2: explain the driver
SPEAKER 1: and I was all day home waiting
SPEAKER 2: he tried to reach you
SPEAKER 1: you know waiting for the driver
SPEAKER 2: he says nobody was home

Fig. 5.3 Challenging segmentation due to interruptions

to identify clear sentence boundaries. For example, consider the following spontaneous utterance in a conversation: "*yeah, you know that movie with the guy, the one who goes like, 'I'll be back' you know.*" In the conversational context, this utterance can be considered a coherent unit conveying a complete thought of trying to remember the title of a famous movie. However, from a grammatical standpoint, it does not constitute a proper sentence with a subject, verb, and object. Instead, it is a string of clauses, fragments, and fillers common in conversational speech. In this case, traditional sentence segmentation methods relying solely on strict grammar rules would struggle to identify an appropriate sentence boundary. Another challenge in transcribing spontaneous speech arises from the presence of interruptions and overlapping speech. Conversations often involve multiple participants engaging in real-time interactions, leading to interruptions where speakers talk over each other. This results in a complex mix of overlapping speech signals that can be challenging to disentangle. Consider the fragment of a dialogue presented in Fig. 5.3.

In this dialogue fragment, the interruptions and overlapping speech complicate the segmentation of individual utterances. To produce sequential utterances, the transcript would need to be artificially transformed by assigning specific segments to each speaker, possibly by re-arranging certain words to form more coherent utterances (Fig. 5.4):

The flexibility and dynamic nature of spontaneous speech present a challenge in accurately segmenting sentences in raw ASR output, requiring the development of specialized techniques that account for the conversational context and linguistic variations. Several techniques can be used to break raw ASR output into segments.

SPEAKER 1: I am sick of this, where is my package?!
SPEAKER 2: Sir, can you please sir just let me explain
SPEAKER 1: It was supposed to be like arriving yesterday
SPEAKER 2: the driver he tried to reach you
SPEAKER 1: and I was all day home waiting you know waiting for the driver
SPEAKER 2: he says nobody was home

Fig. 5.4 Segmented transcript

- *Pause-based segmentation*: Pause-based segmentation of ASR transcripts is a technique that utilizes pauses in speech to identify potential boundaries between sentences or utterances. Pauses serve as natural breaks in speech and can indicate where one thought or idea ends and another begins. Algorithms that employ pause-based segmentation analyze the duration and frequency of pauses in the transcript to identify likely segmentation points. Longer pauses are typically indicative of sentence or utterance endings, while shorter pauses might indicate pauses within a sentence. Pauses can be detected based on conventional energy-based features of the sound signal, but they can also be derived from spectral features [34]. A specific set of methods that utilize pauses for ASR output processing is *Voice Activity Detection* (VAD). The purpose of VAD is to identify segments of the audio that contain speech and filter out non-speech segments. By accurately detecting voice activity, ASR systems can focus on transcribing only the speech segments. VAD algorithms typically analyze various characteristics of the audio signal, such as amplitude, frequency content, and energy, to differentiate speech from background noise or silence. This enables ASR systems to effectively recognize and process speech by ignoring irrelevant or non-vocal parts of the audio. It should be noted, though, that VAD is rather a pre-processing technique which filters out non-speech segments, which is more useful in case of long recordings. In the context of spontaneous conversations such non-speech segments are usually short.
- *Acoustic and prosodic cues*: Acoustic cues (changes in pitch, energy, and duration) and prosody (rhythm, intonation, stress patterns of speech) can provide valuable information about the structure of speech [36]. These cues help identify boundaries between sentences or utterances, as they often exhibit distinct patterns at the end of a thought or idea. For instance, a falling pitch contour or a lengthened final syllable may indicate the end of a sentence.
- *Language modeling*: Language modeling can be used for ASR output segmentation by estimating the likelihood of sentence boundaries based on statistical analysis of textual data [25]. This approach assumes that sentences are constructed based on certain grammatical rules and that there are specific cues in the language that signal sentence boundaries. However, applying language modeling to segment ASR output, especially for spontaneous speech, can be challenging due to the mismatch between spontaneous speech and written language patterns reflected in language model. Language modeling is more suited for the task of *ASR post-processing for readability*, where the aim is to transform the raw ASR output into more readable text while keeping the semantic contents of utterances [24]. Language models can be successfully used to transform an ASR output like "*and I um wanted to uhm to ask if I can like withdraw you know withdraw cash from yhm atm of some you know other bank*" into "*I want to ask if I can withdraw cash from an ATM of a different bank.*"
- *Dependency parsing*: Dependency parsing can help overcome challenges of spontaneous speech segmentation by using the syntactic relationships between words [15]. One common approach to segmenting ASR output into sentences using dependency parsing is to build the dependency tree after raw ASR output has been processed by the punctuation restoration algorithm. A sentence boundary

detection algorithm, such as a rule-based or machine learning-based approach, can be used to identify potential sentence boundaries based on punctuation marks like periods, question marks, and exclamation marks. Initial sentence boundaries can be further reviewed by looking for specific dependency relations that indicate the end of a sentence. If a dependency relation implies that a word is closely connected to the previous sentence's structure, it can be considered part of the previous sentence. On the other hand, if a dependency relation indicates the beginning of a new sentence, a new sentence boundary can be created. Using the syntactic relationships between words, dependency parsing can help identify sentence boundaries more accurately in spontaneous speech transcriptions. It allows for a more sophisticated approach compared to simple punctuation-based methods, which may not handle complex sentence structures well. However, it is important to note that dependency parsing is not flawless, especially in the presence of highly disfluent or non-standard speech.

- *Machine learning and deep learning*: A supervised classification model can be trained to classify each word or punctuation mark in the ASR output as either a sentence boundary or not. The model can be trained on a labeled dataset where each word or punctuation mark is annotated with its corresponding sentence boundary label. Features such as word context, POS tags, or acoustic features can be extracted to represent each word. Examples of supervised machine learning algorithms for sentence boundary detection include Conditional Random Fields, Hidden Markov Models, and Support Vector Machines [3]. The trained model can then be used to predict sentence boundaries in new ASR output. Alternatively, the problem of transcript segmentation can be expressed as the sequence-to-sequence learning task [21, 23]. The ASR output is treated as a sequence of words, and the model is trained to predict the sentence boundaries. The sequence-to-sequence model consists of an encoder that encodes the input sequence (ASR output) and a decoder that generates the predicted sentence boundaries. Attention mechanisms help the model focus on relevant words while making predictions. This approach requires a large annotated corpus or parallel data with sentence boundary labels to train the sequence-to-sequence model effectively. The main downside of both machine learning and deep learning approaches is that they require a large and diverse training dataset to improve the model's ability to generalize.

5.4 True Casing and Punctuation Restoration

The problem of *true casing* in engineering the transcripts of ASR recordings lies in the accuracy of preserving the correct casing of proper names [26]. The lack of true casing in the ASR output can be attributed to the constraints imposed by language models and vocabularies used in ASR systems. These models are designed to recognize and transcribe words based on their frequency and occurrence in the training data. While some very popular proper names like major rivers or mountains may be included in the system's vocabulary, the vast majority of proper names, such as people's names, company names, or organization names, are not explicitly present in the vocabulary. As a consequence, when the ASR system encounters a proper name that is not part of its vocabulary, it attempts to approximate the pronunciation and often fails to determine the appropriate casing. This leads to an error, as the system may treat the name as a regular word or simply use a standard capitalization format. Consequently, the lack of comprehensive coverage in the ASR vocabulary results in inaccurate casing of most proper names within the transcription. Consider the utterance "*I will search this using Google*" which could easily be transcribed by the ASR as "*i will search this using goggle*". In this case, not only does the ASR system produce a word error by transcribing "*Google*" as "*goggle*", but it also introduces a capitalization error by changing it from a proper noun to a lowercase common noun. Even more difficult is proper casing of peoples' names because the diversity of word errors in transcribing peoples' names is boundless. For instance, the utterance "*meet Sarah Boycot*" can be very easily misinterpreted as "*me saw her buy coat.*" As a general rule, sentence initials can be capitalized, but even this depends on the quality of sentence segmentation technique. Very popular proper nouns can be introduced into ASR vocabulary (e.g., to properly capitalize words such as "England", "New York", or "Asia"), but beyond that proper capitalization must be handled by the language model based on the grammatical structure of sentences. The failure of the ASR to properly capitalize words has also downstream consequences. For instance, NER models often utilize casing to infer the beginning and the end of a named entity span. Consider the following output of the ASR: "*meet me in front of the bureau of labor statistics building at five pm*". Most NER models will fail to correctly identify the `ORG` entity span, but when the casing is properly restored to "*meet me in front of the Bureau of Labor Statistics building at five pm*", the entity span is much easier to mark.

Most commercially available ASR systems do not include punctuation or capitalization in their outputs, which poses a significant challenge for downstream NLP tasks. To truly understand spoken language, it is important to comprehend both the meaning and grammar of spoken utterances, which relies on accurate punctuation. Punctuation is crucial for intent annotation. Let us consider the example of annotating instances of negative sentiment in call transcripts. If we have the phrase "*nobody came back and i don't like i said he didn't leave a slip*", it would be incorrectly marked as negative sentiment because of the presence of the utterance "*i don't like*". However, the correct punctuation should be "nobody came back and i don't, like i said, he didn't leave a slip". Missing punctuation can introduce errors in phrases with personal references ("*let's eat grandma!*" vs "*let's eat, grandma!*"), enumerations ("*i love cooking my family and pets*" vs "*i like cooking, my family, and pets*", or prepositions at sentence boundaries ("*taken care of the customer*" vs "*taken care of. The customer.*") This lack of punctuation becomes even more problematic in the presence of ASR errors, which can account for a large portion of transcribed words in spontaneous human conversations. For example, if the ASR mistakenly transcribes *shoes* as *shows*, introducing correct punctuation into the utterance "*i am looking for new shows my old ones are falling apart* " can be quite challenging. For some languages punctuation marks are vital to distinguish between the interrogative and declarative sentences. For instance, in Portuguese, the interrogative sentence "*do you want coffee?*" and the declarative sentence "*you want coffee*" translate to the same "*você quer café*" and can only be distinguished based on intonation and the presence of the question mark at the end of the interrogative sentence. Finally, the task of punctuation restoration is made more challenging due to stochastic ASR errors. For spontaneous human conversations word errors can amount to 15%-20% of transcribed words. Consider the following transcript: "*Hi my name's we will do I have pleasure speaking with today*". It may be very challenging to correctly introduce punctuation since the actual utterance is "*Hi my name is Eve who do I have pleasure speaking with today*". Of course, there are applications where punctuation is not required. For instance, if the task is to index text for future keyword-based information retrieval, punctuation does not provide any benefit. For certain applications punctuation can be introduced explicitly into the transcript. This is the case with dictation, when a person can name a punctuation mark (e.g. by saying aloud "*coma*") and the mentioned punctuation mark will be inserted into the text.

An important, yet often overlooked, aspect of spontaneous human conversations is the overlap of utterances [35]. The overlap in conversational speech is substantial, and it is as frequent in phone conversations between strangers as in face-to-face meetings between close acquaintances. The presence of the overlap primarily makes diarization of speech more challenging as the boundaries between speaking parties blur. More interestingly, though, the presence of overlap changes the distribution of punctuation marks in the transcript. The speakers dynamically manage the dialogue by introducing repetitions, self-corrections, and deletions into the conversation. As the result, questions are much more common in spontaneous dialogues then they are in regular language. Speakers use questions to verify comprehension, to uphold the flow of conversation, to request acknowledgements from the other party, etc. Consequently, punctuation restoration model must be ready to produce much more question marks than an off-the-shelf language model.

The simplest form of punctuation prediction is the discovery of sentence-like unit boundaries, where the problem is the binary classification (with "period" and "space" classes). Historically, many different techniques have been tried, for instance, Word Confusion Matrices [20], Maximum Entropy Models [22], Conditional Random Fields [28], Hidden Markov Models [41], and mixtures of probability models [27]. Features used to detect sentence boundaries included both linguistic information (n-gram language models, turn markers, part of speech annotations) [37] and prosodic features [13, 36, 40]. The advent of modern deep neural networks introduced unprecedented advancements in punctuation prediction. Recurrent neural networks quickly surpassed previous state-of-the-art models. Apart from incorporating word embeddings into punctuation prediction [9], these models initially employed LSTM architectures to predict punctuation marks using longer contexts [38, 39]. This initial work considered only two punctuation marks: commas and periods. All other popular punctuation marks (question marks, exclamation marks, semicolons, colons were mapped to periods and other punctuation marks have been removed from the corpus). The authors have decided to retain only pauses and drop all other prosodic

Table 5.1 The results of an LSTM-based punctuation restoration (*Source* [38])

Class	Precision	Recall	F_1
,	0.71	0.65	0.66
.	0.57	0.61	0.58

features (F0 contours, phoneme durations, energy values) because previous research suggests that pause durations are by far the most useful feature. The duration of a pause does not seem to be a very good predictor for commas, despite the common conviction that people make pauses in places where commas should be. In many languages there are strict grammatical rules about the use of commas, and there are sets of words which are always preceded by a comma. Thus, a bi-directional LSTM can easily spot places, where the comma should be placed based on the following word. Lists of items belonging to the same category also suggest the presence of a comma. There are some indications that topical shifts also increase the probability of a period. Of course, in many languages there are sets of words which typically start a new sentence, so knowing these words in advance helps restore periods better.

Let us look closely at this solution proposed by Tilk and Alumäe [38]. The first model is a simple one-hot encoder followed by a hyperbolic tangent activation which feeds into an LSTM cell. The weighted output of LSTM cell is fed into a softmax layer. The output of the LSTM layer is forwarded to the second model which incorporates word pause durations as input features. The architecture of the second model is identical to the first model. The model learns how to use pause durations along with textual features for predicting punctuation marks. The reason for splitting the training between two models (instead of including prosodic features directly in the first model or adapting existing parameters) is the fact that the corpus with annotated pauses is much smaller than the original textual corpus. It is much faster to train this as a separate model. Secondly, according to the authors, stacking a new classifier on top of the LSTM layer produces better results. The model is compared with two baselines: a hidden event language model and a decision tree using identical set of features. The hidden event language model is a 4-gram language model with the vocabulary of 100K (the same as LSTM). The dataset for training and evaluating the solution is 145.4 hours of audio transcribed into 386M tokens. The texts represent various types of subjects, from newspapers, web, fiction, magazines, to broadcast news and conversations. The broadcast conversations are the smallest part (505K out of 386M tokens) so it is hardly a good representative of spontaneous conversations. The final model reaches the metrics presented in Table 5.1.

As we can see, when the model is applied to ASR output, the results are not very impressive. The errors introduced by the ASR clearly influence and hurt the performance of the model as compared to manually transcribed reference text. Period restoration also seems to be more difficult that comma restoration.

More recently, simpler architectures have proven to be sufficiently robust. Character-level convolutional neural networks (CNNs) can restore punctuation marks efficiently and these models do not suffer from out-of-vocabulary tokens or long inference times. At the same time, CNNs struggle to catch longer contexts necessary to restore question marks [17]. In the last two years, Transformer-based models have been gaining popularity for both general punctuation prediction and more specific tasks, such as question prediction [8] or disfluency removal [10]. The popularity of Transformer-based models can be easily explained by the usefulness of the mechanism of attention in punctuation prediction [31, 39]. Another area of active research is the reframing of the punctuation prediction task in terms of machine translation. These works are mainly driven by real-time translation services [11, 12, 32].

A good example of Transformer-based approach is presented by Nguyen *et al.* in [30] where the authors propose to use Transformer architecture to restore capitalization and punctuation in ASR transcripts with self-attention and chunk merging. The key insight is the fact that punctuation restoration and true casing (capitalization restoration) should be done jointly as true casing uses the same features and also is required for many downstream tasks (e.g. NER). The second insight is that many models divide long input sequences into fix-length non-overlapped chunks before being fed into the model. This may introduce errors around wrongly selected borders of chunks and, as the consequence, bad prediction of words and punctuation marks around chunk's boundary due the lack of sufficient context. The authors propose to use an end-to-end model with four stages: split input sequence into overlapping chunks, perform true casing and punctuation restoration on each chunk, perform chunk merging, and decide what to keep in the output sequence.

Chunking is the act of extracting phrases from unorganized text. This involves analyzing a sentence and identifying its components, such as noun groups, verbs, and verb groups. However, it does not determine their internal structure or their role in the overall sentence. Chunking builds upon POS tagging, using POS-tags as input and producing chunks as output. Usually, chunking breaks down phrases into smaller pieces (still larger than words) with implicit meaning, that are more useful than individual words. It removes all modifiers around the required phrase and leaves only the important words alone from lengthy descriptions. Chunking can be regarded as a form of information extraction. There are two types of chunking: chunking up (extract the overview of the information), and chunking down (extract detailed information from the sentence). This up/down distinction is similar to deductive or inductive conclusion through the use of the language. It allows to reach higher meanings or search for more specific bits of missing information.

Because the authors suggest to use Transformers, the length of the chunk does not have to be very long and the self-attention mechanism works better than earlier solutions like RNNs or LSTMs. LSTM is known to learn from much longer contexts than a Transformer, but for punctuation prediction the context does not have to be that long. The Transformers have 6 hidden layers, word embedding of 256, batch size of 4096 and training for 200 epochs. Each head in the model has 8 heads. The training on 500 MB of data takes 20h on NVIDIA 2080Ti GPU. The authors notice that the models often make mistakes close to chunk borders. The solution is to significantly overlap windows (the experiments have chunks of k tokens with $k/2$ overlap). Unfortunately, the authors experiment on British National Corpus (BNC)

Table 5.2 The results of a Transformer-based punctuation restoration (*Source* [30])

Class	Precision	Recall	F_1
Upper case	0.74	0.53	0.62
,	0.43	0.41	0.42
.	0.10	0.87	0.19
?	0.49	0.22	0.30

and not on a real ASR output (so this experiment is out of domain, the characteristics of input strings are totally different from the output of the ASR). The test set consists of 67K sentences. Training is performed on 6M sentences. There are 4.6M periods and 4.9M commas in the training set. The results are presented in Table 5.2

Essentially, any text with punctuation can be utilized for training a punctuation model. However, the existing textual datasets are often not suitable for representing spontaneous speech. Patterns learned from sources like Wikipedia, Web Crawl, or news corpora do not apply to transcriptions of casual conversations. Creating new datasets with annotations is also a challenging task. The original conversational transcripts lack legibility, and manually adding punctuation marks is a laborious and costly process.

Our solution to this problem is presented in [2]. Our approach is based on the hypothesis that it is possible to reduce stochastic ASR errors by retrofitting static word embeddings specific to the application domain. While ASR errors cannot be rectified during punctuation prediction, retrofitting the word embeddings allows to enhance the accuracy of punctuation prediction models. We validate this hypothesis by demonstrating how pre-trained GloVe embeddings [33] can be retrofitted to the domain of call center conversations using Mittens [16]. This retrofitting process significantly improves the accuracy of punctuation prediction in transcribed calls. Unlike general patterns of punctuation found in corpora used for training GloVe embeddings, spontaneous conversations have distinct punctuation patterns. At the same time, pre-trained GloVe embeddings contain a wealth of general knowledge. By utilizing Mittens, we manage to merge the domain-specific punctuation data from the Fisher corpus with the general knowledge present in pre-trained GloVe embeddings. The effectiveness of the approach is demonstrated using a bi-directional LSTM model and a CNN model with pre-trained embeddings retrofitted to the punctuation prediction task.

We have chosen the Fisher corpus [14] as the primary training data for punctuation restoration model. The creation of the Fisher corpus involved numerous participants, with each participant making a few short calls. These speakers were typically unfamiliar with each other, resulting in a wide range of speaking styles and vocabulary. This diversity of speakers was intended to represent various demographic categories such as gender, age, dialect region, and English fluency. The Fisher corpus exhibits

Table 5.3 The distribution of punctuation classes in the Fisher corpus (*Source* [2])

Class	Count	Percentage (%)
ϵ (blank)	1 429 905	79.1
,	208 289	11.5
.	148 624	8.2
?	22 182	1.2

an imbalanced distribution of punctuation classes, which is common in conversational speech (Table 5.3). This makes the Fisher corpus a good candidate dataset for training and evaluating punctuation prediction in ASR transcripts.

A conversation is represented as a sequence $\mathcal{C} = [w_1, w_2, \ldots, w_n]$ of words. Each word w_i is composed of the following components: t_i, which represents the textual form of the word; c_i, a binary feature indicating which side of the conversation uttered the word; s_i, a real number indicating the starting time of the word; d_i, a real number indicating the duration of the word; and p_i, the punctuation mark that follows the word. The inclusion of mixed conversation sides allows for an effective representation of interjections, interruptions, overlap, and simultaneous speech. The punctuation marks are known only during training and are predicted during inference. The punctuation prediction problem is treated as a sequence tagging task.

We use three types of features. Initially, static word embeddings are utilized. Specifically, we choose 300-dimensional pre-trained GloVe vectors, which have undergone retrofitting. The next set of features consists of the interval (offset) between the start of the current word and the end of the previous word, as well as the duration of the current word. These features are standardized in relation to the other words uttered by the same speaker in the same conversation. Consequently, pauses are not explicitly represented as word tokens, but rather inferred by the model based on the timings (offsets and durations) of the words.

Word embeddings have become a popular approach for transfer learning in language processing. While various methods exist for training these embeddings, the underlying idea is to represent the likelihood of a word occurring in the context of other words through dense vectors. These probabilities are derived from extensive corpora, such as Wikipedia text, product reviews, or social media. The prediction restoration model uses pre-trained GloVe embeddings trained on the Common Crawl dataset. This dataset comprises 2.6 billion textual documents collected from the internet. GloVe trains word representations in such a way that, for a pair of words w_i and w_j, the dot product of their embeddings, $\widehat{w}_i \cdot \widehat{w}_j$, approximates the log-probability of the co-occurrence of these words in the training corpus. Let V denote the available vocabulary and let c_{ij} denote the co-occurrence of words w_i and w_j. The original objective function of GloVe is:

$$J = \sum_{w_i, w_j \in V} f(c_{ij})(\widehat{w}_i^T \widehat{w}_j + b_i + b_j - \log c_{ij})^2 \tag{5.1}$$

where b_i and b_j are the bias terms which represent the frequency of w_i and w_j in the training corpus, respectively. Function $f(c_{ij})$ attenuates the impact of infrequent word pairs, at the same time increasing the impact of frequent word pairs.

To address the issue of language models in health-care data, Dingwall and Potts [16] introduced a retrofitting model for static embeddings. The authors reframe the task of training GloVe embeddings in a matrix formulation and suggest extending the objective function to incorporate domain data. Specifically, they propose retrofitting the original embeddings by penalizing the squared distance against the new vector $\widehat{d_i}$ for all words d_i belonging to the domain vocabulary D. The impact of retrofitting can be controlled using the weight μ. As a result, the original GloVe objective function J is modified to:

$$J_{\text{Mittens}} = J + \mu \sum_{d_i \in D} \|\widehat{w}_i - \widehat{d}_i\|^2. \tag{5.2}$$

Put simply, the revised objective function applies dual influences by simultaneously optimizing the original GloVe objective, which aligns word representations with log probabilities of word-pair occurrences (left term), and enforcing a penalty whenever domain-specific word representation deviates from the original embedding (right term).

In the described scenario, the punctuation used in conversational transcripts noticeably differs from the punctuation utilized in the Common Crawl corpus. The transcripts are prone to ASR errors, where inaccurate transcriptions may contain homophone sequences. However, if these erroneous sequences are brief (consisting of 1-2 words), it is likely that retrofitting can rectify them, as the ASR errors occur in similar contexts as the correct words. Assuming this hypothesis holds true, the approach would help address the challenge of lacking high-quality embeddings trained on domain-specific corpora.

Let us now discuss the experimental evaluation of this mode of punctuation restoration. All the models employed in our study share the same CNN architecture. Each model incorporates multiple layers of 1D convolutions, which can be interpreted as fully-connected layers that process input within small windows. Following each layer, a SELU activation function is applied, which yielded a slight improvement compared to batch normalization with ReLU. Various combinations of layer numbers, filter sizes, and other hyper-parameters were tested during experimentation. The most successful and consistent results were obtained using six 1D convolutional layers with a filter size of 128 and zero padding. Kernel sizes of 3 were used for all layers except the last one, which had a kernel size of 20. Notably, these kernels were capable of incorporating words from both sides of the conversation. The hidden layers all utilized a dilation rate set to 2. Model regularization techniques were also explored. Firstly, a dropout of 0.5 was added before the softmax layer. Secondly, weight decay (L2 with a weight of 0.001) was implemented for the softmax layer. Gaussian noise with a standard deviation of $\sigma = 0.1$ was introduced before the final softmax activation and SELU activations. Finally, by employing SELU activations, the weights were constrained to adhere to a normal distribution with the mean of

$\mu = 0$ and the standard deviation of $\sigma = 1$. The last layer of the model consisted of a fully-connected layer with softmax activation. It was separately applied to each time step to predict the appropriate punctuation for a given word. Categorical cross-entropy was used as the loss function. Learning rate was adaptively decreased by the factor of 0.5 with the patience set to 3 epochs. The minimum learning rate was set to $\eta = 1 \times 10^{-5}$ and the batch size was $bs = 256$.

Three variants of CNN-based models were compared with CNN and BiLSTM baselines in the study conducted by Zelasko et al. [42]. The BiLSTM model was composed of four BiLSTM layers, each containing 128 weights in both directions. The CNN model consisted of six layers of 1D convolutions, followed by SELU activations. The convolutional layers utilized filters with a length of 128 characters and the dilation rate set to 2. The context was specified as three words, except for the last layer which utilized a context of 20 words. Both models incorporated a fully connected last layer, followed by a dropout layer with a rate of 0.50, and a softmax layer. A weight decay of 0.001 was applied to the softmax layer. Gaussian noise with a standard deviation of $\sigma = 0.1$ was introduced to distort the inputs, time feature, and softmax activation. Evaluation of each model involved calculating precision, recall, and F_1 score for each individual punctuation class. The detailed results can be found in Table 5.4.

We have chosen a convolutional neural network and standard pre-trained GloVe embeddings for the baseline. We consider this to be a strong baseline, as it has been proven to be a viable solution for a production-ready, real-time system according to Zelasko [42]. We have constrained the selection of the 50 000 words to only those that appear in the ASR vocabulary, creating the CNN-50k model. This model covers words that appear in the ASR vocabulary twice as frequently as the baseline model and it achieves approximately 60% coverage of the vocabulary. We note that there is an improvement in both precision and recall across all punctuation classes. The final model, which has embeddings that cover almost the entire ASR vocabulary (with around 20% missing from GloVe), outperforms the baseline in all three metrics considered. It does this without significantly increasing resource requirements or incurring additional computational costs. Finally, we compare our CNN-based model with a BiLSTM model based on GloVe embeddings. Our model outperforms the baseline in terms of precision and recall for all punctuation classes, except for full stops. One additional advantage of using a CNN-based model in the production environment is that it can be parallelized more easily compared to the BiLSTM architecture.

Table 5.5 presents a comparison of the confusion matrices for the CNN-baseline and the CNN-100k-mittens models. The latter model improves the accuracy of all three punctuation classes. The most significant improvement is seen in the "period" class, with a 9% absolute improvement. The "question mark" and "comma" classes also show improvements of 6.2% and 6.8% respectively. We attribute these improvements to the fact that the CNN-100k-mittens model predicts punctuation marks that were missed by the baseline model. In other words, our model inserts punctuation marks in places where the baseline model predicted blanks.

Table 5.4 The per-class precision, recall, and F_1 score (in %) achieved by the compared models on the pre-trained GloVe embeddings with additional retrofitting using Mittens (*Source* [2])

Model	Class	Precision	Recall	F1
CNN-baseline	ϵ	92.7	**95.8**	94.2
	.	65.5	58.7	61.9
	?	67.5	49.0	56.8
	,	66.6	55.1	60.3
BiLSTM-baseline	ϵ	***93.5***	94.7	94.1
	.	67.9	66.7	67.3
	?	64.7	54.6	59.2
	,	68.2	**64.1**	**66.1**
CNN-50k	ϵ	92.6	95.5	94.0
	.	70.2	65.0	67.5
	?	70.2	51.7	59.5
	,	69.7	60.8	65.0
CNN-50k-mittens	ϵ	93.3	95.3	94.3
	.	70.7	**68.7**	69.7
	?	**72.8**	53.7	61.8
	,	69.3	62.4	65.7
CNN-100k-mittens	ϵ	93.1	95.6	**94.3**
	.	**71.8**	67.7	**69.7**
	?	71.2	**55.2**	**62.2**
	,	**69.9**	61.9	65.7

Table 5.5 Confusion matrix comparison for the CNN-baseline and CNN-100k-mittens models (*Source* [2])

	CNN baseline				CNN 100k mittens			
	ϵ	.	?	,	ϵ	.	?	,
ϵ	95.8	1.4	0.2	2.6	95.6	1.1	0.1	2.8
.	32.2	58.7	1.5	7.6	21.2	67.7	1.5	9.6
?	23.7	21.0	49.0	6.3	19.9	18.4	55.2	6.4
,	37.8	6.5	0.6	55.0	31.1	6.6	0.4	61.9

In order to observe the effects of retrofitting word embeddings on homophones, we have calculated the cosine similarity between certain pairs of confusing utterances in both the original and retrofitted embedded space. We see that the vectors are much closer to each other in the retrofitted space compared to the original GloVe space (Table 5.6).

Punctuation prediction plays an important role in enhancing the readability and segmentation of transcripts generated by ASR. However, ASR introduces inherent

Table 5.6 Similarity between homophones in the original GloVe space and in the retrofitted space (*Source* [2])

Homophone	Cosine similarity	
	Original	Retrofitted
I have to cancel I have to cancer	0.74	0.91
By Buy	0.16	0.4
Go to the court Got the card	0.69	0.78
Thank you Think you	0.84	0.9

errors in word recognition and results in a significant variation in the distribution of punctuation marks due to the nature of spontaneous speech. These factors restrict the effectiveness of word embeddings because most static word embeddings are trained using correctly segmented sentences from written text. It has been hypothesized that aligning pre-trained word embeddings with the word co-occurrence structure observed in transcribed calls could help overcome some of the challenges posed by ASR and enhance the accuracy of punctuation prediction. For this purpose, we have used the recently published technique called Mittens to retrofit the GloVe embeddings using call transcripts obtained from ASR. The results of the retrofitting process showed a consistent improvement of 6%–9% across all punctuation classes compared to the original GloVe embeddings. Additionally, the use of retrofitted embeddings allowed to replace the BiLSTM model by a faster CNN-100k-mittens model. This practical finding is significant, as CNN-based models are more suitable for deployment in production environments due to their adherence to inference time constraints.

5.5 Closing

In this chapter, our focus shifted towards transcript engineering. We explored the problem of transcript alignment and annotation, looking at techniques that enable accurate synchronization of audio and text. We discussed the importance of segmentation, which allows to break down transcripts into meaningful units, facilitating further analysis. True casing and punctuation restoration played important roles in improving the quality and readability of the transcripts.

Throughout this chapter, it became evident that the challenge of transcript engineering remains unresolved. The fundamental issue is the stark dissimilarity between spontaneous speech and written text, attributing to the misalignment between NLP tools trained on prescriptive language and the language encountered in ASR output. It is reasonable to assert that NLP tools operating on ASR transcripts persistently operate in the out-of-domain mode, constantly struggling with adaptation. In the next chapter we will look at the downstream NLP tasks and we will examine how they are impacted by the characteristics of spontaneous speech artifacts.

References

1. J. Allen, M. Core, Draft of damsl: Dialog act markup in several layers (1997)
2. Ł. Augustyniak, P. Szymanski, M. Morzy, P. Zelasko, A. Szymczak, J. Mizgajski, Y. Carmiel, N. Dehak, Punctuation prediction in spontaneous conversations: Can we mitigate asr errors with retrofitted word embeddings? arXiv preprint arXiv:2004.05985 (2020)
3. S. Brognaux, T. Drugman, Hmm-based speech segmentation: Improvements of fully automatic approaches. IEEE/ACM Trans. Audio Speech Lang. Process. **24**(1), 5–15 (2015)
4. H. Bunt, Dynamic interpretation and dialogue theory. Struct. Multimodal Dial. **2**, 139–166 (1999)
5. H. Bunt, Dimensions in dialogue act annotation. In: *LREC*, pp. 919–924 (2006)
6. H. Bunt, The dit++ taxonomy for functional dialogue markup, in AAMAS, Workshop. Towards a Standard Markup Language for Embodied Dialogue Acts **2009**, 13–24 (2009)
7. H. Bunt, J. Alexandersson, J. Carletta, J.-W. Choe, A.C. Fang, K. Lee, V. Petukhova, A. Popescu-Belis, L. Romary, C. Soria, et al., Towards an iso standard for dialogue act annotation, in *Seventh conference on International Language Resources and Evaluation (LREC'10)* (2010)
8. Y. Cai, D. Wang, Question mark prediction by BERT, in *Asia-Pacific Signal and Information Processing Association Annual Summit and Conference* (IEEE, 2019), pp. 363–367
9. X. Che, C. Wang, H. Yang, C. Meinel, Punctuation prediction for unsegmented transcript based on word vector, in *10th International Conference on Language Resources and Evaluation* (2016), pp. 654–658
10. Q. Chen, M. Chen, B. Li, W. Wang, Controllable time-delay transformer for real-time punctuation prediction and disfluency detection. arXiv preprint arXiv:2003.01309 (2020)
11. E. Cho, J. Niehues, K. Kilgour, A. Waibel, Punctuation insertion for real-time spoken language translation, in *11th International Workshop on Spoken Language Translation* (2015)
12. E. Cho, J. Niehues, A. Waibel, NMT-based segmentation and punctuation insertion for real-time spoken language translation, in *Interspeech* (2017), pp. 2645–2649
13. H. Christensen, Y. Gotoh, S. Renals, Punctuation annotation using statistical prosody models, in *ISCA Tutorial and Research Workshop (ITRW) on Prosody in Speech Recognition and Understanding* (2001)
14. C. Cieri, D. Miller, K. Walker, The Fisher corpus: a resource for the next generations of speech-to-text. Int. Conf. Lang. Resour. Eval. **4**, 69–71 (2004)
15. S. Davidson, D. Yu, Z. Yu, Dependency parsing for spoken dialog systems. arXiv preprint arXiv:1909.03317 (2019)
16. N. Dingwall, C. Potts, Mittens: an extension of glove for learning domain-specialized representations. arXiv preprint arXiv:1803.09901 (2018)
17. W. Gale, S. Parthasarathy, Experiments in character-level neural network models for punctuation, in *Interspeech* (2017), pp. 2794–2798
18. A. Haubold, J.R. Kender, Alignment of speech to highly imperfect text transcriptions, in *2007 IEEE International Conference on Multimedia and Expo* (IEEE, 2007), pp. 224–227
19. T.J. Hazen, Automatic alignment and error correction of human generated transcripts for long speech recordings, in *Ninth International Conference on Spoken Language Processing* (2006)
20. D. Hillard, M. Ostendorf, A. Stolcke, Y. Liu, E. Shriberg, Improving automatic sentence boundary detection with confusion networks. Technical report, Washington University (2004)
21. W.J. Hou, Y.S. Guo, L.-R. Dai, Segment boundary detection directed attention for online end-to-end speech recognition. EURASIP J. Audio Speech Music Process. **1–16**, 2020 (2020)

22. J. Huang, G. Zweig, Maximum entropy model for punctuation annotation from speech, in *7th International Conference on Spoken Language Processing* (2002)
23. W.R. Huang, S.-Y. Chang, D. Rybach, R. Prabhavalkar, T.N. Sainath, C. Allauzen, C. Peyser, Z. Lu, E2e segmenter: Joint segmenting and decoding for long-form asr. arXiv preprint arXiv:2204.10749 (2022)
24. J. Liao, S. Eskimez, L. Liyang, Y. Shi, M. Gong, L. Shou, Q. Hong, M. Zeng, Improving readability for automatic speech recognition transcription. ACM Trans. Asian Low-Resour. Lang. Inf. Process. **22**(5), 1–23 (2023)
25. J. Liao, Y. Shi, M. Gong, L. Shou, S. Eskimez, L. Lu, H. Qu, M. Zeng, Generating human readable transcript for automatic speech recognition with pre-trained language model, in *ICASSP 2021-2021 IEEE International Conference on Acoustics, Speech and Signal Processing (ICASSP)* (IEEE, 2021), pp. 7578–7582
26. L.V. Lita, A. Ittycheriah, S. Roukos, N. Kambhatla, Truecasing, in *Proceedings of the 41st Annual Meeting of the Association for Computational Linguistics* (2003), pp. 152–159
27. Y. Liu, A. Stolcke, E. Shriberg, M. Harper, Comparing and combining generative and posterior probability models: Some advances in sentence boundary detection in speech, in *Conference on Empirical Methods in Natural Language Processing* (2004), pp. 64–71
28. W. Lu, H.T. Ng, Better punctuation prediction with dynamic conditional random fields, in *Conference on Empirical Methods in Natural Language Processing* (Association for Computational Linguistics, 2010), pp. 177–186
29. P.J. Moreno, C.F. Joerg, J.-M. Van Thong, O. Glickman, A recursive algorithm for the forced alignment of very long audio segments. ICSLP **98**, 2711–2714 (1998)
30. B. Nguyen, V.B. Hung Nguyen, H. Nguyen, P.N. Phuong, T.-L. Nguyen, Q.T. Do, L.C. Mai, Fast and accurate capitalization and punctuation for automatic speech recognition using transformer and chunk merging. arXiv preprint arXiv:1908.02404 (2019)
31. A. Mireia Farrús, L. Wanner, Attentional parallel RNNs for generating punctuation in transcribed speech, in *International Conference on Statistical Language and Speech Processing* (Springer, 2017), pp. 131–142
32. S. Peitz, M. Freitag, A. Mauser, H. Ney, Modeling punctuation prediction as machine translation, in *International Workshop on Spoken Language Translation* (2011)
33. J. Pennington, R. Socher, C. Manning, GloVe: Global vectors for word representation, in *Conference on Empirical Methods in Natural Language Processing* (2014), pp. 1532–1543
34. J.-L. Shen, J.-W. Hung, L.-S. Lee, Robust entropy-based endpoint detection for speech recognition in noisy environments, in *Fifth international conference on spoken language processing* (1998)
35. E. Shriberg, A. Stolcke, D. Baron, Observations on overlap: findings and implications for automatic processing of multi-party conversation, in *Interspeech* (Citeseer, 2001), pp. 1359–1362
36. E. Shriberg, A. Stolcke, D. Hakkani-Tür, G. Tür, Prosody-based automatic segmentation of speech into sentences and topics. Speech Commun. **32**(1–2), 127–154 (2000)
37. A. Stolcke, E. Shriberg, Automatic linguistic segmentation of conversational speech, in *4th International Conference on Spoken Language Processing*, vol. 2 (IEEE, 1996), pp. 1005–1008
38. O. Tilk, T. Alumäe, LSTM for punctuation restoration in speech transcripts, in *16th Annual Conference of the International Speech Communication Association* (2015)
39. O. Tilk, T. Alumae, Bidirectional recurrent neural network with attention mechanism for punctuation restoration, in *Interspeech* (2016), pp. 3047–3051
40. D. Wang, S.S. Narayanan, A multi-pass linear fold algorithm for sentence boundary detection using prosodic cues, in *IEEE International Conference on Acoustics, Speech, and Signal Processing*, vol. 1 (IEEE, 2004), pp. I–525
41. J.P. Yamron, I. Carp, L. Gillick, S. Lowe, P. van Mulbregt, A hidden markov model approach to text segmentation and event tracking, in *IEEE International Conference on Acoustics, Speech and Signal Processing*, vol. 1 (IEEE, 1998), pp. 333–336
42. P. Żelasko, P. Szymański, J. Mizgajski, A. Szymczak, Y. Carmiel, N. Dehak, Punctuation prediction model for conversational speech, in *Interspeech* (2018), pp. 2633–2637

Chapter 6
Transcript Understanding

6.1 Intent Discovery

Spoken language understanding consists in identifying and processing of semantically meaningful parts of the dialogue, most often performed on the transcript of the dialogue produced by the ASR system. These meaningful parts are referred to as *dialog acts* (see Sect. 3.3.1) and provide structure to the flow of conversation. But the recognition and classification of dialog acts is not sufficient for true spoken language understanding. Each dialog act can be instantiated to form an *intent*. Intents are simply sets of utterances which exemplify the intention of the speaking party to perform a certain action, convey or obtain information, express opinion, etc. For instance, the intent *Account Check* can be expressed using examples such as "*let me go over your account*", "*found account under your name*", "*i have found your account*". At the same time, the intent *Account Check* would be an instance of the dialog act *Statement*. In other words, an intent represents what the speaker intends to communicate or achieve through their words. Understanding the intent of a conversation is crucial for effective communication and meaningful interaction. It allows participants to interpret the meaning behind the words and respond appropriately. Intents can vary widely and may include expressing an opinion, asking for information, making a request, providing clarification, expressing emotions, or any other communicative purpose. Analyzing and interpreting the intents of different speakers can help in building better rapport, resolving conflicts, and overall improving the quality of conversation. Examples of typical intents encountered in call center conversations include customer requests (*Refund*, *Purchase Intent*), desired actions by an agent (*Up-selling*, *Order Confirmation*), and customer satisfaction risks (*Customer Service Complaint*, *Supervisor Escalation*).

Rescoring the transcript using sentences that represent given intents can be an effective approach to correct ASR mistranslations. Intents, which reflect the expected purpose or goal of a statement or question in a conversation, can serve as reliable indicators of what should occur at a specific point in the transcript. By comparing

M. Morzy, *Spoken Language Processing*, Studies in Computational Intelligence 1205,
https://doi.org/10.1007/978-3-031-88566-2_6

the decoded sentence from the ASR with the expected intents, it becomes possible to identify instances where an intent is not accurately conveyed due to ASR errors. If a very similar sentence is decoded by the ASR, but fails to incorporate the expected intent, rescoring the transcript allows for corrections to be made. This process ensures that the ASR output aligns more closely with the intended meaning, enhancing the overall accuracy and fidelity of the transcribed conversation. As an example, consider the intent *End of Hold* which represents an utterance that should be pronounced by the call center agent on the return to the conversation with a client, after keeping the client on hold. If the ASR decodes the utterance as "*thank you for your patients*", based on the intent definition, we can correct it to "*thank you for your patience*" even if this hypothesis was not the most probable ASR hypothesis for the utterance. Other downstream applications of intent discovery include conversation monitoring (e.g., real-time obfuscation of personally identifiable information) or live assistance in conversation (e.g., next best action recommendation).

Defining intents can be a challenging and costly process for several reasons. Firstly, each intent represents a piece of domain knowledge, requiring input from subject matter experts who have a deep understanding of the specific field or industry. This expertise is necessary to accurately capture the nuances of the intent and ensure its relevance within the given domain. Next, there is a significant variability in how intents can be expressed in speech. Different individuals may phrase the same intent in various ways, adding complexity to the intent definition process. Additionally, the presence of named entities within intents further complicates matters. Some phrases may require the use of placeholders to represent the dynamic nature of named entities in conversations. Consider the intent *Problem Statement* which can be instantiated using phrases such as "*I can't get my printer to print*", or "*My dishwasher does not start*". In order to make this intent more general, the definitions of the intent should not be bound to specific devices. For instance, an example of the intent could be "*I can't make my* `__DEVICE__` to", "*the* `__DEVICE__` doesn't start", or "*I have a problem with my* `__DEVICE__`". In these examples, the token `__DEVICE__` serves as the placeholder for a particular name of the device, brand name, or model name. The algorithm of intent discovery must allow for such placeholders, preferably allowing also for typing the placeholders. Again, let us consider a hypothetical intent *Delay Complaint* which represents the complaint of a customer related to the delay in service, delivery, or answer. An example of this intent could be the utterance "*I have waited for* `__DELAY_TIME__`", but only if `__DELAY_TIME__` masks a token which is indeed an instance of `TIME`, `INTERVAL`, or `NUMBER` entities. So the utterance "*I have waited for three days*" would be an instance of the intent, whereas the utterance "*I have waited for my wife*" would not. Similarly, with entity templating an example "*your flight departs* `__SYSTEM_TIME__`" would match both "*your flight departs* in ten minutes" and "*your flight departs* tomorrow at seven forty five pm". This relieves the annotator from enumerating millions of possible examples for each entity and facilitates the creation of more specific examples that increase precision. To illustrate, "*your item number is*" could incorrectly match "*your item number is wrong*", but "*your item number is* `__SYSTEM_NUMBER__`" would not.

The collection of intent definitions primarily relies on the manual work of data annotators. These annotators analyze and label the intent instances in the data, which can be a time-consuming and labor-intensive task. Unfortunately, only a small portion of intents can be transferred between different domains. While some intents, like *Call Opening* or *Customer Verification* may have a broader applicability, most intents are domain-specific. For instance, intents such as *Question: Account Lookup* or *Question: Refund Viability* need to be redefined and tailored to each new domain, requiring additional time and effort. In order to overcome the high cost of entering a new domain, one has to introduce a method for rapid domain adaptation based on a large library of intent definitions. Let us now briefly discuss one possible architecture that can be applied to transcripts with the use of intent definitions. The method utilizes tools such as Word Confusion Networks, Finite State Transducers, and ASR lattices, which all have been discussed in Sect. 4.3.5. The method has been developed primarily by Jan Mizgajski and Piotr Żelasko [26].

The solution begins with the annotation tool. Defining all examples by hand would be prohibitively expensive and cause intents to have limited recall and precision, as, by virtue of combinatorial complexity of language, each intent needs hundreds of examples. To alleviate this problem the annotators should be supported by a set of tools, including: fast transcript annotation interface for initial discovery of intents, an interactive system for semi-automatic generation of examples which recommends synonyms and matches examples on existing transcripts for validation, and unsupervised or semi-supervised methods based on sentence similarity and grammatical pattern search for new intent and examples discovery.

In addition, the notion of an example must be extended with two concepts that improve the recall of a single example:

- *Blank quota*, which defines the number of words that may be found *in-between* the words of the example and still be acceptable, e.g., "*this is* very *outrageous*" becomes a potential match for "*this is outrageous*" if the blank quota is greater than 0. This allows the annotator to focus on the words that convey the meaning of the phrase and ignore potential filler words.
- *Entity templating* allowing examples to incorporate entities in their definitions.

The above methods allow the annotators to create hundreds of intents efficiently, with tens of thousands of examples, matching millions of *distinct* potential phrases. When combined with the ability to configure entities and select a subset of intents that are relevant to a particular domain, this approach produces highly customer-specific repositories of domain knowledge.

Now, let us present the algorithm for intent discovery. The lattice $\mathcal{L}$ is an acceptor, where each arc contains a symbol representing a single word in the current hypothesis (see Fig. 6.1). The working assumption is the closed ASR vocabulary and FSTs operate on word-level, rather than character- or phoneme-level. Note that this assumption is not a limitation of the method. Should the ASR have an unlimited vocabulary (as some end-to-end ASR systems do), it is possible to dynamically construct the lattice symbol table and merge it with the symbol table of intent definitions.

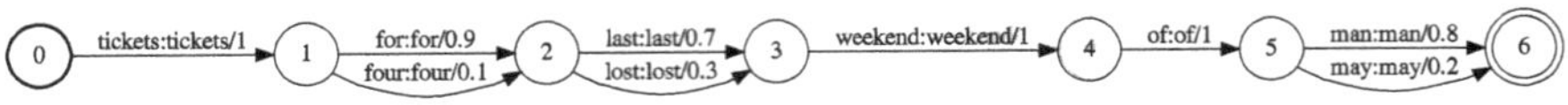

Fig. 6.1 Word confusion network representing the lattice $\mathcal{L}$ (source: [53])

To perform intent annotation (i.e., to recognize and mark the position of intent instances in the transcript), first the FST index $\mathcal{I}$ is created with all intent examples. This index is a transducer which maps the alphabet of words (input symbols) onto the alphabet of intents (output symbols). The index $\mathcal{I}$ is constructed in such a way that its composition with the lattice results in another transducer $\mathcal{A} = \mathcal{L} \circ \mathcal{I}$ representing the annotated lattice. The method begins by creating a single FST state which serves as both the initial and the final state and contains a single loop wildcard arc. A wildcard arc accepts any input symbol and transduces it to an empty ϵ output symbol. The wildcard arc can be efficiently implemented with special σ-matchers, available in the OpenFST framework [1]. Composition with the singleton FST maps every input symbol in $\mathcal{L}$ to ϵ, which denotes the lack of intent annotations. For each intent example, an additional branch (i.e. a set of paths) is created in the index $\mathcal{I}$ which maps multiple sequences of words to a set of symbols representing this particular intent example (see Fig. 6.2).

The method uses three types of symbols: an intent symbol ι (including begin ι_B, continuation ι_C and end ι_E symbols), an entity symbol ω (including an entity placeholder symbol ω^*), and a null symbol ϵ. The intent symbol ι is the delimiter of the intent annotation and it demarcates the words constituting the intent. The begin (ι_B) and continuation (ι_C) symbols are mapped onto arcs with words as input symbols, and the end (ι_E) symbol is inserted in an additional arc with an ϵ input symbol after the annotated fragment of text. It is important that the begin symbol ι_B does not introduce an extra input of ϵ. Otherwise, the FST composition is inefficient, as it tries to enter this path on every arc in the lattice $\mathcal{L}$. The entity symbol ω marks the presence of an entity in the intent annotation. Each entity in the intent index $\mathcal{I}$ is constructed as a non-terminal entity placeholder ω^*, which allows using the FST lazy replacement algorithm to enter a separate FST grammar $\mathcal{G}$ describing a set of possible values for a given entity. When it is possible to provide a comprehensive list of entity instances, the transducer can be used, otherwise, an approximation of this list can be obtained by running a named entity recognition model predictions on an n-best list (see Fig. 6.3). Finally, the null symbol ϵ means that either no intent was recognized, or the word spanned by the intent annotation did not contribute to the annotation itself.

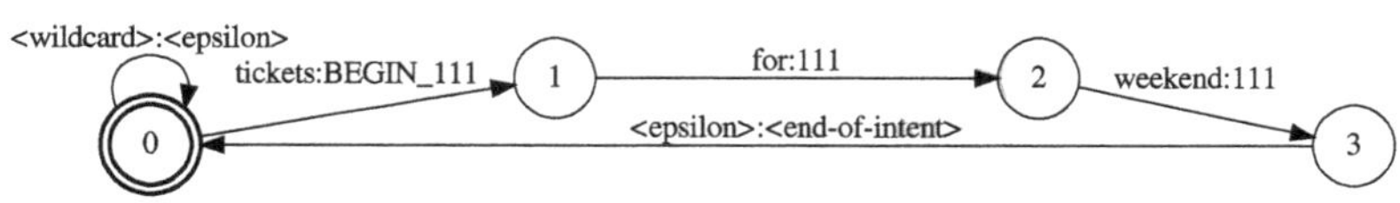

Fig. 6.2 An index $\mathcal{I}$ matching a single intent example *tickets for weekend* to an intent number 111 (source: [53])

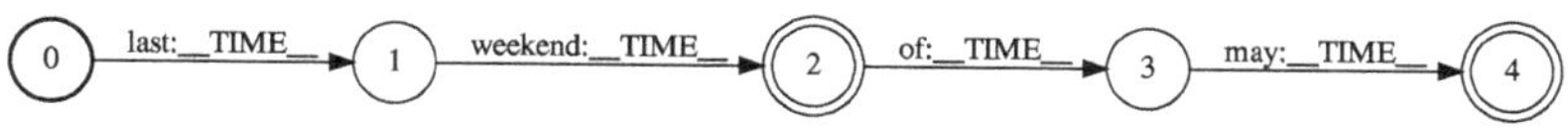

Fig. 6.3 A simple grammar FST $\mathcal{E}$ for the non-terminal token *__TIME__*. Note that both states 2 and 4 are final (indicated by the double circle) (source: [53])

This procedure successfully performs exact matching of the transcription to intent index, when all words present in the current lattice path are also found in the intent example. Unfortunately, this approach is highly impractical when real transcriptions are considered. Sequences of significant words are interwoven with filler phonemes and words, for instance the utterance "*I want uhm to order like um three yyh three tickets*" could not be matched with the intent example "*I want to order __NUMBER__ tickets*". To overcome this limitation the intent index $\mathcal{I}$ has to be adapted to enable fuzzy matching so that some number n of filler words can be inserted between words constituting an intent example, while still allowing to match the intent annotation. The wildcard arcs between each of the intent-matching words provide the matching capacity of 0 to n matches of any word in the alphabet. The example of such an index is shown in Fig. 6.4.

The naive implementation allowing for n superfluous (non-intent) words appearing within intent-matching words would lead to a significant explosion of the annotations spans. Instead, a post-processing filtering step can be employed to prune lattice paths where the number of the allowed non-intent word is exceeded. The filtering step has a computational complexity of $O(|S| + |A|)$, where $|S|$ is the number of states and $|A|$ is the number of arcs in the non-pruned annotated lattice $\mathcal{L}'$. The pruning algorithm is based on the depth-first search (DFS) traversal of the lattice $\mathcal{L}'$ and marks each state in $\mathcal{L}'$ as either new, visited, or pruned. Only new states are entered and each state is entered at most once. The FST arcs are only marked as either visited or pruned. Each FST state keeps track of whether an intent annotation parsing has begun (i.e., a begin symbol ι_B has been encountered but the end symbol ι_E has not appeared yet) and how many wildcard words have been matched so far.

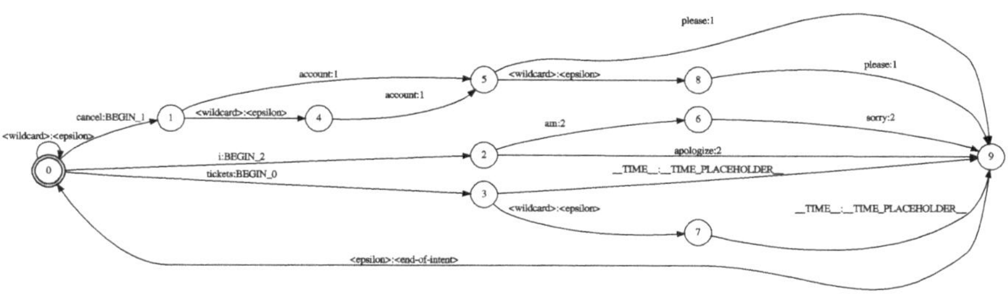

Fig. 6.4 The intent index $\mathcal{I}$ which matches three different intent examples: *cancel account please* with a blank quota of 1, *i apologize* with the synonym *am sorry*, and *tickets __SYSTEM_TIME__*, where the last token is a special non-terminal token, replaced dynamically during composition with an appropriate grammar FST (source: [53])

The traversal of the lattice $\mathcal{L}'$ is stateful. It starts in a *non-matching* state and remains in this state until encountering the intent begin symbol ι_B. Then the state becomes *matching* and remains such until encountering the intent end symbol ι_E. A state is marked as *pruned* when the traversal state is *matching* and the number of wildcard words exceeds the blank quota for the given intent example. Any arc incident with a pruned state is not entered during further traversal, leading to a significant speed-up of lattice processing. After every possible path in the lattice $\mathcal{L}'$ has been either traversed or pruned, all redundant (i.e., pruned or not visited) FST states are removed from the lattice, along with all incident arcs. The annotated lattice $\mathcal{A}$ is obtained after final traversal of the lattice $\mathcal{L}'$ which prunes arcs representing unmatched word alternatives. If no intent has been matched on any of the parallel arcs, the traversal retains only the best path hypothesis. Figure 6.5 presents an example of the annotated lattice before and after pruning.

Despite significant pruning described in the previous section, the annotated lattice $\mathcal{A}$ still contains competing variants of the transcript. The next step consists in selecting the "best" variant and traversing all paths in $\mathcal{A}$ which correspond to this transcript. The key concept of the method is to guide the selection of the "best" variant by intents rather than word probabilities. The likelihood of a particular longer sequence of words in the language is lower than the likelihood of a particular shorter sequence of words. Since *a priori* longer intent examples are less likely to appear in the lattice $\mathcal{A}$ purely by chance, the presence of a lattice path containing a longer intent example provides strong evidence for that path. The complete set of heuristics applied sequentially to the annotated lattice $\mathcal{A}$ in search of the best path is the following:

(a) select the path with the longest intent annotation,
(b) select the path with the largest number of intent annotations,

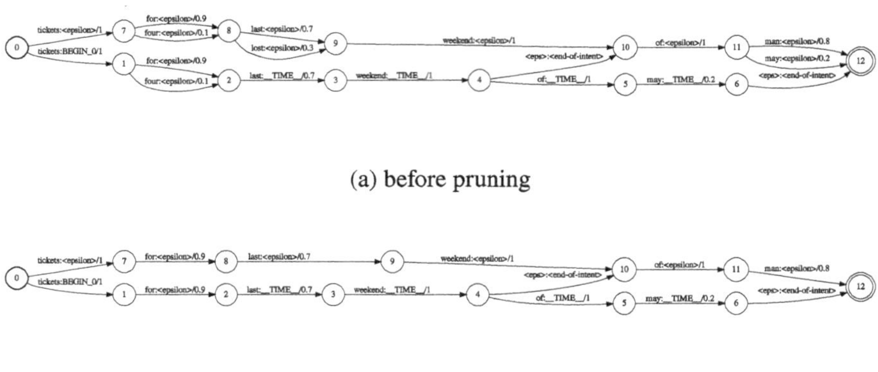

(a) before pruning

(b) after pruning

Fig. 6.5 Annotated lattice $\mathcal{A}$ resulting from composition $\mathcal{L}' \circ \mathcal{I}$ with replacement using $\mathcal{E}$ before **a** and after **b** pruning. Note that the last word *man* was rescored as *may* in the path due to the recognition of an annotated intent (source: [53])

Table 6.1 Rescoring accuracy w.r.t. intent length measured on the annotated dataset of 500 rescored intents (source: [53])

Intent length	Occurrences	Accuracy [%]
1	25	32.0
2	139	39.5
3	149	63.7
4	80	76.2
5	53	94.3
6	19	94.7
7+	35	100.0

(c) select the path with the intent annotation with the longest span (i.e. consider also blank words),
(d) select the path with the highest original ASR likelihood.

The chosen best path is composed with the annotated lattice $\mathcal{A}$ to produce the annotated lattice $\mathcal{A}^*$ with the final variant of the transcript. The output intent annotations are retrieved by traversing every path in $\mathcal{A}^*$.

The entire procedure presented above suffers from a serious drawback. As mentioned before, most NLP tasks are performed on the turn level, which naturally corresponds to the turn-taking patterns in a typical dialogue. This approach yields good results for chatbot conversational interfaces or information retrieval systems, but for spontaneous human-to-human dialogues, the demarcation of turns is much more difficult. Thus, we cannot expect those intent examples would align with ASR segments which capture a single speaker turn. This issue can be addressed by concatenating turn-level lattices of all utterances of a person throughout the conversation into a conversation-lattice $\mathcal{L}^C$. This lattice can still be effectively annotated and pruned using algorithms presented above to obtain the annotated conversation-lattice $\mathcal{A}^C$.

Unfortunately, the annotated conversation-lattice $\mathcal{A}^C$ cannot be parsed in search of the best path using the algorithm presented previously, because the computational cost of every path traversal in $\mathcal{A}^C$ is exponential in the number of words. Fortunately, one can exploit the structure of the conversation-lattice $\mathcal{A}^C$ to identify the best path. The lattice $\mathcal{A}^C$ is a sequence of segments organized either in series or in parallel. Segments with no intent annotations are series of linear word hypotheses, which branch to parallel word hypotheses whenever an intent annotation is matched (because the original path with no intent annotation is retained in the lattice). The parallel segment ends with the end of the intent annotation. These series and parallel segments can be detected by inspecting the cumulative sum of the difference of out-degree and in-degree of each state in a topologically sorted conversation-lattice $\mathcal{A}^C$. For series regions, this sum will be equal to 0, and greater than 0 in parallel regions. The computational cost of performing this segmentation is $O(|S| + |A|)$, i.e., linear in the number of states and arcs in the annotated conversation-lattice $\mathcal{A}^C$. After having performed the segmentation, the partial best path search in parallel segments is resolved using the presented method.

How effective are the intent definitions in overcoming ASR errors and correcting transcripts? Below we present the quantitative analysis of the proposed algorithm with respect to a baseline which annotates only the best ASR hypothesis (see Table 6.1). The experiments are performed with an intent library comprised of 313 intents in total, each of which is expressed using 170 examples on average. The annotations are performed on more than 70 000 US English phone conversations with an average duration of 11 minutes, but some of them take even over one hour. The topics of these conversations span across several domains, such as inquiry for account information or instructions, refund requests or service cancellations. Each domain uses a relevant subset of the intent library (typically 100–150 intents are active). To evaluate the effectiveness of the proposed algorithm, a dataset of 500 intent annotations has been sampled. These intent annotations were found in the lattices due to rescoring in the cancellations and refunds domain. The correctness of the rescoring was judged by two annotators, who labeled 250 examples each, and verified by another annotator in the second pass. The annotators read the whole conversation transcript and listened to the recording to establish whether the rescoring is meaningful. In cases when a rescored word was close but technically incorrect, e.g., a mistake between the present and past tense of a verb, and led to the recognition of a correct intent, the intent annotation was labeled as correct.

The longer the intent annotation, the more likely it is to be correct due to stronger contextuality of the annotation. Intent annotations which span at least three words are more likely to rescore the lattice correctly than to introduce a false positive. These results also lead to a practical heuristic, that an intent annotation which spans only one or two words should not be considered for rescoring. Application of this heuristic results in an estimated accuracy of 77%. A stricter heuristic requires at least four words span and yields an accuracy of 87.7%. Calibration of this threshold is helpful when the algorithm is adapted to a downstream task, which may either require higher recall or precision of the recognized intent annotations. Some examples of successful lattice rescoring are presented in Table 6.2. Note how "*acme that calm*" has been rescored to "*acme dot com*", or how "*think of a collie level*", which is nonsensical, is rescored to "*thank you for calling*".

The proposed algorithm finds 658 549 intents in all conversations, covering 4.1% of all (62 450 768) words, whereas baseline algorithm finds 526 356 intents, covering 3.3% of all words. Therefore, the increase in intent recognition of the method is 25.1% by rescoring 8.3% of all annotated words (0.34% of all words). Particular intents achieve different improvements ranging from no improvement up to 1062%—aggregated percentile results are presented in Table 6.3. We see that half of all intents gained at least 35.7% of improvement, while 20% of all intents gained at least 83.5%.

Table 6.2 Examples of successful lattice rescoring along with the recognized intent. In first example real brand name was obfuscated by a fictional brand name ACME. Words in parentheses are turns of another speaker (source: [53])

Intent	Text
Website Mention	Original: just use your regular acme (ok) that calm pay with credit card Rescored: just use your regular acme (ok) dot com pay with credit card
Account Lookup	Original: can you looked at my count Rescored: can you look at my account
Account Information	Original: i need to recount (mhm sure) number or email addresses Rescored: i need your account (mhm sure) number or email address
Call Opening	Think of a collie level Rescored: thank you for calling
Refund	Original: work connie the refined Rescored: work on the refund

Table 6.3 Cumulative lift distribution of improvement in intent recognition (source: [53])

Percentile	Minimum improvement [%]
10	128.9
20	83.5
30	62.4
40	49.4
50	35.7
60	28.7
70	21.6
80	13.7
90	2.0

6.2 Named Entity Recognition

An *entity* is a specific word or phrase that represents a unique and identifiable information in the conversation or can be assigned to a category with well defined semantics. The former leads to the concept of a *named entity*, where a token represents some real world object, such as a location (*New York*), a person (*Lionel Messi*), a brand (*Apple*). The latter can represent concepts such as money (*one hundred dollars*), date (*first of May*), duration (*two hours*), credit card number, etc. Named entities play an important role in NLP and information extraction tasks. By identifying and categorizing named entities, it becomes possible to extract meaningful information from the conversation, such as identifying key individuals or organizations mentioned, understanding geographical references, or extracting important dates or events. Named entity recognition (NER) is a common task in NLP and it is essential for various applications, including information retrieval, question answering systems, sentiment analysis, and more. A particular set of labels available to the NER model is called the *labeling scheme*, and typical labeling schemes include entity classes such as `PER` (person), `GPE` (geopolitical entity), `ORG` (organization), `LOC` (locations), and entities representing types of quantities (`TIME`, `MONEY`, `QUANTITY`, `DATE`, etc.). Some labeling schemes include extended entity types, such as `EVENT`, `PRODUCT`, `WORK_OF_ART`, `NORP` (nationality, religion, political affiliation), or `LAW`. Historically, many types of machine learning models have been used for NER, such as Support Vector Machines [18], Hidden Markov Models [27], and Conditional Random Fields [50]. The seminal work of Nadeau and Sekine [28] surveys a broad spectrum of techniques, starting from rule-based and statistical approaches and culminating with early machine learning techniques. Subsequent works of Marrero et al. [23], Saju and Shaja [33] and Dai [10] extend NER surveys to new domains and complex overlapping entities. More recent surveys, such as the work by Yadav and Bethard [52], Goyal et al. [14] and Li et al. [21] focus on deep learning models for named entity recognition. To the best of author's knowledge, Hatmi et al. [17] were the first to incorporate named entity recognition into the automatic speech transcription process. The authors tagged the ASR dictionary with named entity tags (since ASR cannot produce any words not present in its dictionary). This initial approach has been superseded by methods aiming at training end-to-end joint models for ASR and NER, as proposed by Ghannay et al. [13], Serdyuk et al. [35], and Stiefel and Vu [38]. The authors train ASR systems to predict both transcription tokens and their part-of-speech or named entity tags in these works. An interesting early attempt at NER recognition in speech transcripts has been made by Parada *et al.* [31]. The authors significantly improve their baseline NER model by injecting into transcript features indicative of out-of-vocabulary (OOV) terms. Since many instances of named entities are OOV tokens, this additional information allows the NER model to more precisely spot transcript segments where named entities may appear.

Work presented in this chapter considers only English resources. Annotated datasets with entity spans in transcripts of spontaneous speech have been published for other languages. Favre *et al.* [12] describe the dataset used in the ESTER Named Entity evaluation program for French. Tomashenko *et al.* [43] present the *MEDIA* dataset with dialogue history in French. Another example of a French dataset is *ESTER 2* introduced by Hatmi *et al.* [17]. Named entity recognition from speech has also been a part of the REPERE challenge [19]. Two large datasets with annotated entity spans for Chinese have been recently released: *CNERTA* [39] and *AISHELL-NER* [6]. Annotated transcripts of speech have been added to the SLUE (Spoken Language Understanding) benchmark [36]. The benchmark contains the *Vox Populi* dataset [46] and parts of the *OntoNotes v5* dataset.

6.2.1 *Fine-Tuning NER for Spontaneous Speech*

Despite recent advancements in NLP technology, spontaneous speech still remains a challenging domain for contemporary NLP models. Some researchers go as far as to announce the existence of the so-called *ASR-NLP gap*, which is expressed in the significant degradation of the performance of NLP models on transcripts of spontaneous speech. The main reason why NLP models trained on prescriptive written text perform poorly on the transcripts of spontaneous speech is the fact that unscripted conversations do not share many characteristics of written text [37]. This is particularly consequential for the task of named entity recognition. The grammatical structure of spontaneous speech is very different from written language on which traditional NER models are trained. Grammatical structure is the backbone of NER models, which use syntactic information to predict entity spans. Part-of-speech tags and dependency trees are often used to improve the accuracy of NER models. In case of spontaneous speech, many of these syntactic hints are missing from the transcript. Ill-defined sentence boundaries, phatic expressions, turn-taking artifacts, and other aspects of spontaneous speech confuse NER models trained on written language. And the problem only gets worse because transcripts produced by an ASR do not contain proper sentence segmentation, true-casing of words, or punctuation. Instead, ASR introduces a significant number of errors in the form of insertions, substitutions, or deletions.

The NLP research community has an overly optimistic view of the WERs introduced by ASR systems. Recent experiments show that WERs in transcripts of spontaneous human speech are much higher than expected. As we have mentioned in Sect. 4.1, Szymański *et al.* [42] showed that a transcript of a standard GSM phone call conversation is subject to a 16%-20% error rate. Del Rio *et al.* [11] confirm this result and report how WERs differ between different types of entity spans. Spans related to date, time, and ordinal numbers were observed to have a lower WER than entities related to proper names. Facility names, organizations, and personal names demonstrate a very high WER of 30%–50%. These findings are in stark contrast to initial reports. For instance, Surdeanu *et al.* [40] reported named entity recognition in Switchboard corpus to be within 5% from a system evaluated on clean textual data. Similarly, Béchet *et al.* [4] claim to have achieved approximately 0.90 F_1 for recognizing phone numbers and 0.70 F_1 for recognizing money mentions in the transcripts from the AT&T *How may I help you?* system under 27.4% WER ratio. Favre *et al.* [12] apply NER models to French corpora and achieve 0.74 F_1 for a relatively broad set of named entities.

Precision, recall, and F_1 scores are standard metrics for reporting NER model performance in NLP. However, these metrics can produce unreliable scores when entity spans are marked on spontaneous human conversation transcripts due to the presence of conversational artifacts. To account for the presence of these artifacts, Message Understanding Conference (MUC) [15, 28] introduced metrics that allow for partial matching of an entity span. MUC defines six categories of partial matching based on the degree of span overlap, the type of the matched entity, and the strictness of expectations, as outlined by Batista [3]. Recently, this problem has been addressed by Caubrière *et al.* [5] who argue for the use of slot error rates.

Let us consider an example of an utterance marked with entity spans by a NER model: "*[Bill F. Smith]*$^{\texttt{PERSON}}$ *joined the [LakeShore]*$^{\texttt{ORG}}$ *corporation in [May of 2020]*$^{\texttt{DATE}}$." When the utterance is correctly transcribed, and all tokens are restored to their correct casing, the task of marking entity spans seems to be straightforward. In reality, the input stream to the NER model will be very different from the idealized example presented above. Firstly, the transcript will be riddled with ASR errors. Secondly, the spontaneous speech will contain repetitions, filler words, hesitation markers, back-channeling, and other artifacts. A more probable example of the same utterance would be "*bill eh [smith]*$^{\texttt{PERSON}}$ *joined uhm lake sure corporation in man [twenty twenty]*$^{\texttt{CARDINAL}}$". The combination of ASR errors, lack of true casing, lack of normalization of dates, and additions of spontaneous speech, make the same NLP model[1] to miss half of the `PERSON` entity span, miss the entire `ORG` entity span, and replace the `DATE` with the `CARDINAL` entity.

In the field of traditional NLP, one technique to improve the performance of statistical models is domain fine-tuning. A large model pre-trained on a well-defined and well-understood dataset can be "pushed" in the direction of the domain of application. Fine-tuning usually involves presenting the model with data different from the data the model has been initially trained on, combined with hyper-parameter optimization or changing the objective function. The assumption behind fine-tuning is that there exists some universal parameter space covering a wide spectrum of tasks, and the process of fine-tuning shifts the model in the parameter space in the direction of the desired task. In other words, fine-tuning assumes that adding a dataset provides more information to the model about the language structure with respect to a given language task. In general, this assumption is met in the case of written texts which share well-defined grammatical structure. It is not obvious, however, if this assumption holds for spoken language.

The main question is the following: is there a convergence of patterns characteristic of named entity mentions in spontaneous speech across different application domains? In other words, are there common features of how people talk about named entities in spontaneous conversations, independent of the domain of conversation? If the answer to this question is "yes," then we can expect that fine-tuning NER models using transcripts with entity span annotations from various domains should improve the performance of those models across other domains. In practical terms, we want to know if it is more beneficial to train a NER model from scratch on a smaller dataset, albeit well aligned with the application domain, or if it is better to fine-tune a pre-trained model with as many NER-annotated transcripts as possible. Here we present the results of our research led by Szymański [41].

The state-of-the-art models for NER are most often evaluated on two datasets, the *CoNLL-2003/CoNLL++* [34], and the *OntoNotes v5* [49]. In the evaluation of the efficacy of fine-tuning the *CoNLL-2003/CoNLL++* dataset has been omitted because

[1] In both examples the `en_core_web_sm` v.3.5.0 model from the `spaCy` library has been used to mark entity spans.

the labeling scheme for this dataset contains only four different classes. In all practical applications, the set of entity classes identified in the conversational transcript is much larger. The next choice involves the datasets for model fine-tuning. Ideally, these would be datasets with transcripts of spontaneous conversations annotated with entity spans. Unfortunately, curating such datasets is extremely expensive, and only a few such datasets exist. Most datasets used for training ASR models, such as the *Fisher* dataset [9], the *LibriSpeech* [30], or the *Common Voice* dataset [2] do not have entity span annotations and do not represent spontaneous speech. In practice, there are only three datasets suitable for fine-tuning NER models for spontaneous speech. The datasets listed below are the only publicly available datasets with both entity spans annotated in transcripts and audio segments for ASR training.

- *OntoNotes v5* (O): the LDC-released OntoNotes v5 [49] with texts from news, broadcast/telephone conversations, and Web data annotated with 18 entity types.
- *SwitchBoard Named Entities* (S): data from Switchboard Dialog Acts Corpus annotated with entity tags following the OntoNotes v5 annotation scheme [7].
- *Earnings21* (E): audio and transcriptions of 44 public phone calls which span almost 40 h of recordings of spontaneous human conversations, with 25 different entity classes annotated in transcripts [11].

The outline of the evaluation is the following. Two models serve as the baseline: `roberta-base` and `roberta-large` [22] trained from scratch. These baseline models are compared with `tner-xlm-roberta-base-ontonotes5` and `tner-xlm-roberta-large-ontonotes5` [44]. The latter are models with the same architecture as baseline models but pre-trained on the *OntoNotes v5* dataset. Each model is fine-tuned using *Earnings21*, *SWNE*, or both datasets. Finally, each combination is tested on all three datasets for each entity class. Thus, there are 36 strategies for selecting the training dataset, the fine-tuning dataset, and the testing dataset.

Each dataset is represented as a normalized gold transcription. The normalization includes the restoration of the true casing of words, the restoration of correct punctuation, and ensuring a coherent spelling of acronyms and abbreviations (e.g., the transcript contains "*IBM*" instead of "*i b m*"). The datasets differ in the way the numbers are normalized. The *Earnings21* transcripts contain numbers written as digits, whereas *SWNE* contains a mixture of digits and words (e.g., the transcript may contain "*two thousand and twenty*" instead of "*2020*"). It is worth stressing the fact that the experimental setting is already biased in favor of NER models. The quality of gold transcripts is much higher than the quality of transcripts produced by ASRs. Standard ASR output is lower-cased without punctuation, and the ASR performs the output segmentation into tokens. The restoration of true casing in spontaneous speech transcripts is very hard, existing models struggle even with punctuation restoration, as we have shown in Sect. 5.4.

Let us begin by presenting the average F_1 score across all entity classes for each strategy of NER model training (see Table 6.4). The first conclusion is that models pre-trained on the *OntoNotes v5* dataset perform competitively only when evaluated on the same dataset but fail to generalize to other domains. The smallest pre-trained model (`base+O`) lags behind the same model fine-tuned for both *Earnings21* and *SwitchBoard*. When tested on *Earnings21*, the model pre-trained on *OntoNotes v5* has $F_1 = 0.51$, when model fine-tuned on *Earnings21* and *SwitchBoard* have $F_1 = 0.81$ and $F_1 = 0.67$, respectively. The larger version `large+O` performs even worse ($F_1 = 045$) when `large+E` has $F_1 = 0.81$ and `large+S` has $F_1 = 0.67$. The same scenario is repeated when models are tested on *SwitchBoard*. Again, model versions fine-tuned on *SwitchBoard* perform best, followed closely by models fine-tuned on *Earnings21*. Interestingly, fine-tuning on both *Earnings21* and *SwitchBoard* is never beneficial. The models fine-tuned with both datasets do not perform better than models fine-tuned with one dataset aligned with the domain. On the other hand, models pre-trained on *OntoNotes v5* profit from extensive fine-tuning but only in the sense of fixing the performance degeneration introduced by pre-training. For instance, `base+O+S+E` has $F_1 = 0.79$ on *Earnings21* and $F_1 = 0.81$ on *SwitchBoard*, which is significantly better than `base+O` ($F_1 = 0.51$ and $F_1 = 0.44$, respectively), but `base+E` has $F_1 = 0.81$ on *Earnings21* and `base+S` has $F_1 = 0.84$ on *SwitchBoard*.

The results presented in Table 6.4 allow us to draw the following conclusions:

- Pre-training on *OntoNotes v5*, despite being a popular choice for NER models, has a detrimental effect and significantly lowers the F_1-score for other domains. Training a model from scratch with domain fine-tuning is always the preferred strategy.

Table 6.4 Average F_1 scores for all strategies of NER model training; E-Earnings21, O-OntoNotes v5, S-SwitchBoard Named Entities (source: [41])

	F_1		
Strategy	E	O	S
base+O	0.51	0.71	0.44
base+S	0.67	0.68	0.84
base+E	**0.81**	0.73	0.75
base+O+S	0.65	0.72	0.84
base+O+E	0.79	**0.76**	0.70
base+S+E	0.80	0.72	0.84
base+O+S+E	0.79	**0.76**	0.81
large+O	0.45	0.58	0.41
large+S	0.67	0.68	**0.86**
large+E	**0.81**	0.70	0.72
large+O+S	0.63	0.72	0.85
large+O+E	0.80	**0.76**	0.71
large+S+E	**0.81**	0.70	0.84
large+O+S+E	0.80	0.75	0.82

Table 6.5 F_1 scores for the entity `CARDINAL`; O-OntoNotes v5, S-SwitchBoard Named Entities, E-Earnings21 (source: [41])

Model	Test	Fine-tuning						
		O	S	E	O+S	O+E	S+E	O+S+E
Base	O	**0.78**	0.64	*0.69*	0.66	0.69	0.69	0.68
	S	0.52	0.83	*0.77*	**0.84**	0.75	0.82	0.82
	E	0.51	*0.74*	**0.81**	0.73	0.77	0.80	0.79
Large	O	**0.73**	0.62	*0.69*	0.71	0.71	0.67	0.69
	S	0.49	**0.83**	0.75	0.76	*0.76*	0.82	0.82
	E	0.51	0.67	**0.81**	*0.80*	0.80	0.80	0.79

- No significant improvement is observed when `roberta-base` is replaced with `roberta-large`. Fine-tuning has a much stronger effect than the size of the model. The additional cost of training a larger model may not be offset by the improved performance.
- Extensive fine-tuning is not beneficial. Models perform better when fine-tuned on a single in-domain dataset. Adding more data to fine-tuning does not result in improved performance.

Let us now scrutinize the detailed results obtained for specific entity classes. In the following tables, boldface is used to mark strategies with the overall highest F_1 for each validation dataset, and italics are used to mark best-performing strategies where the validation dataset has not been used for fine-tuning. Table 6.5 presents the results for the `CARDINAL` class. In spoken language, instances of this class can be easily mistaken for utterances referring to time, money, or dates. As we can see, the best strategies result in very similar F_1 scores across all validation datasets ($0.78 - 0.84$), with `roberta-base` slightly outperforming `roberta-large`. The differences between in-domain training (validation performed on the dataset used for fine-tuning) and out-of-domain training (validation performed on the dataset not used for fine-tuning) are less than 0.10 of the F_1 score.

Now, let us compare the results for the `CARDINAL` entity class with the results for the `DATE` class presented in Table 6.6. Dates in spontaneous speech suffer from the problem of normalization. For instance, the year 2023 can be uttered as "*two thousand and twenty three*", "*twenty twenty three*", or even "*twenty three*" depending on the context. The way this utterance will be reflected in the transcript depends on the particular model employed by the ASR, it could be "*2023*", "*20 23*", "*two thousand 23*", or even "*to thousand twenty tree*" (if ASR errors are to be included). The in-domain fine-tuning produces F_1 scores in the range of $0.79 - 0.84$, very similar to the `CARDINAL` class. The main difference is the significant drop in F_1 score for out-of-domain fine-tuning ($0.51 - 0.74$) resulting, most probably, from the differences in date normalization across datasets.

The `base+E` model achieves $F_1 = 0.58$, and the `large+E` model tested on the same dataset deteriorates to $F_1 = 0.49$. When tested on *Earnings21*, both models

Table 6.6 F_1 scores for the entity `DATE`; O-OntoNotes v5, S-SwitchBoard Named Entities, E-Earnings21 (source: [41])

Model	Test	Fine-tuning						
		O	S	E	O+S	O+E	S+E	O+S+E
Base	O	**0.79**	0.61	*0.73*	0.63	0.76	0.71	0.73
	S	0.25	**0.81**	*0.58*	0.80	0.50	0.79	0.76
	E	*0.68*	0.51	0.83	0.49	**0.84**	0.82	0.82
Large	O	**0.78**	0.59	*0.74*	0.60	0.76	0.71	0.75
	S	0.26	**0.83**	0.49	0.81	*0.51*	0.79	0.75
	E	*0.68*	0.44	**0.83**	0.43	**0.83**	0.82	**0.83**

achieve a reasonable $F_1 = 0.83$. When the models are switched to test `base+S` and `large+S` on *Earnings21*, the results are $F_1 = 0.51$ and $F_1 = 0.44$, respectively (even though these models on *SwitchBoard* have $F_1 = 0.81$ and $F_1 = 0.83$, respectively). In other words, models fine-tuned on one domain perform noticeably worse on other domains, and enriching them with pre-training on *OntoNotes v5* does not help. Interestingly, this phenomenon is most pronounced for entity classes that should be well-defined (one can see similar behavior for `TIME`). The hypothesis is that this behavior is caused by the differences in transcript normalization. It seems that the Transformer architecture struggles with overcoming normalizations. Anecdotally, `roberta-large` can be tested by computing sentence similarity for the anchor ("*seventeenth of may twenty twenty seven*") and for three variations ("*May 17th, 2027*", 0.672), ("*17th of May 2027*", 0.737), and ("*seventeenth of may two thousand twenty seven*," 0.895). What is surprising is that all these expressions convey very similar information, yet the similarity score varies significantly depending on the normalization scheme. A similar phenomenon may take place in NER models based on the Transformer architecture.

Another entity class that is highly dependent on normalization and which is ambiguous in spontaneous speech is the `TIME` class. Table 6.7 presents the results for this class. Except for the models fine-tuned and validated on the *SwitchBoard* datasets, the performance of all other strategies is in the range of $0.57 - 0.62$ for in-domain training and $0.53 - 0.68$ for out-of-domain training. This result suggests that the `TIME` entity spans are generally difficult to mark in spontaneous conversation transcripts, and fine-tuning is not helpful in this task.

Finally, let us consider the `PRODUCT` entity class which is very domain-dependent. Obviously, what constitutes a product depends on the domain of application. If the NER model processes customer calls to the call center of a bank, the products being discussed will be credit cards, account types, loans, mortgages, etc. If the transcripts contain customer calls to a travel agency, product spans will contain names of hotels, destinations, packages, insurances, etc. It is, therefore, reasonable that the out-of-domain fine-tuning should not be beneficial in the case of the `PRODUCT` entity. Table 6.8 presents the detailed results. The average results are quite discouraging,

Table 6.7 F_1 scores for the entity `TIME`; O-OntoNotes v5, S-SwitchBoard Named Entities, E-Earnings21 (source: [41])

Model	Test	Fine-tuning						
		O	S	E	O+S	O+E	S+E	O+S+E
Base	O	0.50	0.51	*0.57*	0.51	**0.57**	0.53	0.55
	S	0.32	0.80	*0.66*	**0.82**	0.57	0.78	0.77
	E	0.40	*0.54*	**0.60**	0.53	0.59	**0.60**	0.57
Large	O	0.49	0.46	*0.53*	0.50	**0.61**	*0.53*	0.60
	S	0.30	**0.80**	*0.68*	0.77	0.55	0.78	0.70
	E	0.49	*0.54*	0.61	0.52	**0.62**	0.60	**0.62**

Table 6.8 F_1 scores for the entity `PRODUCT`; O-OntoNotes v5, S-SwitchBoard Named Entities, E-Earnings21 (source: [41])

Model	Test	Fine-tuning						
		O	S	E	O+S	O+E	S+E	O+S+E
Base	O	**0.52**	0.42	*0.47*	0.47	0.49	0.43	**0.52**
	S	0.13	0.24	*0.69*	0.66	0.55	**0.70**	0.64
	E	0.04	**0.78**	0.39	0.23	037	0.40	0.36
Large	O	0.36	0.42	0.42	**0.52**	**0.52**	*0.47*	0.50
	S	0.11	**0.83**	*0.78*	0.64	0.57	0.77	0.69
	E	0.05	*0.30*	0.42	0.21	0.40	0.43	**0.44**

with most strategies resulting in F_1 score close to 0.50. In many scenarios, the recognition of product entity spans falls below 0.30, which makes the NER model useless in practical applications. The average difference between in-domain and out-of-domain training is small, mostly because the performance of models is unsatisfactory.

Table 6.9 presents the detailed results for all entity classes. These results further corroborate presented conclusions. Pre-trained models perform significantly worse than fine-tuned models for each entity class. If the model is fine-tuned with domain data, the performance improves substantially. As expected, entities that are semantically more general are relatively easy to predict (`MONEY`, `PERSON`, `GPE`), while other classes pose a challenge (`PRODUCT`, `EVENT`). There is no experimental evaluation supporting fine-tuning with the diversified dataset, fine-tuning with domain-related data is always the soundest strategy.

There is one more possible reason for low performance of NER models on transcripts of spontaneous speech. Named entities can be referred to implicitly, without being named directly, through contextual cues and inferencing. For example, a speaker may refer to "*my wife*" without explicitly mentioning the person's name and assuming that this name is known to the second party through the context of the conversation. Although this hypothesis lacks support in experimental data, author's long experience with spontaneous speech transcripts suggests that humans do assume the

Table 6.9 Detailed F_1 scores for all strategies of NER model training for each entity class; datasets used for fine-tuning and testing are marked as O-OntoNotes v5, S-SwitchBoard Named Entities, E-Earnings21; averaged results per strategy are marked with A: macro average, B: micro average, C: weighted average (source: [41])

Train	Test	CARD	DATE	EVENT	FAC	GPE	LANG	LAW	LOC	MON	NORP	ORD	ORG	PCT	PER	PROD	QTY	TIME	WoA	A	B	C
Base+S	S	0.83	0.81	0.68	0.24	0.95	1.00	0.00	0.70	0.87	0.87	0.00	0.80	0.00	0.92	0.78	0.79	0.80	0.93	0.75	0.84	0.84
	E	0.74	0.51	0.57	0.17	0.73	0.00	0.11	0.55	0.91	0.72	0.19	0.59	0.95	0.77	0.24	0.44	0.54	0.03	0.49	0.67	0.67
	O	0.64	0.61	0.27	0.24	0.81	0.68	0.06	0.36	0.44	0.70	0.03	0.67	0.45	0.86	0.42	0.49	0.51	0.35	0.48	0.68	0.68
Base+E	S	0.77	0.58	0.78	0.46	0.92	0.73	0.00	0.69	0.81	0.86	0.00	0.76	0.00	0.85	0.69	0.72	0.66	0.83	0.62	0.75	0.75
	E	0.81	0.83	0.60	0.19	0.82	0.00	0.24	0.73	0.92	0.78	0.89	0.70	0.97	0.82	0.39	0.64	0.60	0.00	0.61	0.81	0.81
	O	0.69	0.73	0.32	0.35	0.87	0.52	0.29	0.38	0.45	0.76	0.76	0.72	0.36	0.82	0.47	0.51	0.57	0.36	0.55	0.72	0.73
Base+S+E	S	0.82	0.79	0.72	0.33	0.96	0.93	0.00	0.70	0.89	0.88	0.00	0.80	0.00	0.91	0.70	0.81	0.78	0.88	0.66	0.83	0.84
	E	0.80	0.82	0.63	0.16	0.82	0.00	0.23	0.76	0.92	0.79	0.89	0.69	0.96	0.82	0.40	0.61	0.60	0.00	0.61	0.80	0.80
	O	0.69	0.71	0.32	0.24	0.85	0.62	0.18	0.40	0.45	0.76	0.68	0.72	0.38	0.82	0.43	0.50	0.53	0.35	0.53	0.70	0.72
Large+S	S	0.83	0.83	0.82	0.45	0.95	0.93	0.33	0.73	0.92	0.89	0.00	0.85	0.00	0.90	0.83	0.85	0.80	0.89	0.80	0.86	0.86
	E	0.67	0.44	0.64	0.07	0.73	0.50	0.11	0.50	0.91	0.77	0.14	0.64	0.93	0.83	0.33	0.46	0.54	0.00	0.51	0.67	0.67
	O	0.62	0.59	0.37	0.24	0.83	0.72	0.20	0.32	0.44	0.78	0.02	0.65	0.44	0.85	0.42	0.42	0.46	0.35	0.48	0.68	0.68
Large+E	S	0.75	0.49	0.74	0.40	0.88	1.00	0.00	0.62	0.83	0.87	0.00	0.71	0.00	0.85	0.78	0.72	0.68	0.80	0.62	0.71	0.72
	E	0.81	0.83	0.57	0.10	0.80	0.00	0.40	0.72	0.92	0.74	0.89	0.72	0.96	0.83	0.42	0.69	0.61	0.00	0.61	0.81	0.81
	O	0.69	0.74	0.26	0.48	0.83	0.36	0.14	0.41	0.45	0.77	0.76	0.68	0.34	0.78	0.42	0.50	0.53	0.33	0.53	0.69	0.70
Large+S+E	S	0.82	0.79	0.80	0.43	0.94	1.00	0.33	0.71	0.94	0.87	0.00	0.82	0.00	0.88	0.77	0.83	0.79	0.89	0.70	0.84	0.84
	E	0.80	0.82	0.60	0.15	0.79	0.00	0.33	0.74	0.92	0.75	0.88	0.72	0.96	0.83	0.43	0.70	0.62	0.00	0.61	0.81	0.81
	O	0.67	0.71	0.22	0.44	0.81	0.52	0.13	0.45	0.45	0.75	0.68	0.68	0.32	0.82	0.47	0.54	0.54	0.36	0.53	0.69	0.70
Base+O+S	S	0.84	0.80	0.74	0.36	0.93	0.93	0.00	0.68	0.96	0.90	0.00	0.81	0.00	0.83	0.66	0.88	0.82	0.91	0.75	0.84	0.84
	E	0.73	0.49	0.55	0.04	0.75	0.00	0.13	0.55	0.90	0.76	0.09	0.60	0.93	0.81	0.23	0.37	0.53	0.00	0.47	0.65	0.65
	O	0.66	0.63	0.33	0.21	0.84	0.54	0.29	0.39	0.41	0.84	0.06	0.69	0.70	0.89	0.47	0.52	0.51	0.48	0.52	0.72	0.72

(continued)

Table 6.9 (continued)

Train	Test	CARD	DATE	EVENT	FAC	GPE	LANG	LAW	LOC	MON	NORP	ORD	ORG	PCT	PER	PROD	QTY	TIME	WoA	A	B	C
Base+O+E	S	0.75	0.50	0.57	0.19	0.87	1.00	0.00	0.58	0.83	0.91	0.00	0.73	0.00	0.78	0.55	0.70	0.57	0.78	0.57	0.70	0.70
	E	0.77	0.84	0.58	0.05	0.83	0.00	0.36	0.73	0.91	0.79	0.90	0.69	0.94	0.82	0.37	0.61	0.59	0.00	0.60	0.79	0.79
	O	0.69	0.76	0.49	0.44	0.88	0.31	0.25	0.51	0.43	0.83	0.80	0.74	0.41	0.88	0.49	0.59	0.57	0.48	0.59	0.75	0.76
Base+O+S+E	S	0.82	0.76	0.73	0.43	0.89	0.93	0.00	0.62	0.91	0.91	0.00	0.80	0.00	0.80	0.64	0.90	0.77	0.91	0.66	0.81	0.81
	E	0.79	0.82	0.59	0.16	0.81	0.00	0.33	0.70	0.91	0.78	0.90	0.68	0.94	0.80	0.36	0.58	0.57	0.00	0.59	0.79	0.79
	O	0.68	0.73	0.40	0.26	0.87	0.38	0.16	0.42	0.43	0.83	0.74	0.75	0.49	0.88	0.52	0.58	0.55	0.47	0.56	0.75	0.76
Large+O+S	S	0.83	0.81	0.84	0.44	0.93	0.88	0.29	0.70	0.90	0.92	0.00	0.86	0.00	0.92	0.64	0.82	0.77	0.92	0.78	0.85	0.85
	E	0.65	0.43	0.60	0.18	0.74	0.00	0.08	0.53	0.88	0.74	0.01	0.64	0.91	0.81	0.21	0.42	0.52	0.03	0.47	0.63	0.63
	O	0.67	0.60	0.28	0.40	0.84	0.54	0.25	0.40	0.42	0.83	0.00	0.68	0.62	0.89	0.52	0.47	0.50	0.47	0.52	0.71	0.72
Large+O+E	S	0.76	0.51	0.47	0.24	0.86	0.93	0.00	0.54	0.85	0.91	0.00	0.76	0.00	0.82	0.57	0.70	0.55	0.77	0.57	0.70	0.71
	E	0.80	0.83	0.60	0.33	0.82	0.00	0.23	0.75	0.91	0.78	0.90	0.70	0.95	0.82	0.40	0.60	0.62	0.29	0.63	0.80	0.80
	O	0.71	0.76	0.40	0.55	0.88	0.36	0.18	0.54	0.38	0.84	0.77	0.74	0.32	0.87	0.52	0.66	0.61	0.42	0.59	0.75	0.76
Large+O+S+E	S	0.82	0.75	0.68	0.31	0.94	0.93	0.00	0.67	0.93	0.94	0.00	0.80	0.00	0.88	0.69	0.85	0.70	0.88	0.65	0.82	0.82
	E	0.79	0.83	0.57	0.15	0.83	0.00	0.33	0.76	0.90	0.81	0.91	0.72	0.95	0.82	0.44	0.64	0.62	0.00	0.61	0.80	0.80
	O	0.69	0.75	0.35	0.47	0.86	0.67	0.38	0.57	0.40	0.85	0.71	0.74	0.23	0.88	0.50	0.63	0.60	0.36	0.59	0.74	0.75
Base+O	S	0.52	0.25	0.21	0.35	0.71	0.82	0.00	0.43	0.41	0.63	0.00	0.40	0.00	0.66	0.13	0.33	0.32	0.12	0.35	0.44	0.44
	E	0.51	0.68	0.08	0.18	0.71	0.50	0.02	0.74	0.13	0.64	0.87	0.41	0.33	0.74	0.04	0.46	0.40	0.02	0.41	0.52	0.51
	O	0.78	0.79	0.43	0.39	0.72	0.59	0.37	0.54	0.65	0.70	0.84	0.64	0.58	0.84	0.52	0.62	0.50	0.21	0.60	0.71	0.71
Large+O	S	0.49	0.26	0.18	0.44	0.64	1.00	0.00	0.33	0.38	0.53	0.00	0.38	0.00	0.62	0.11	0.31	0.33	0.19	0.34	0.41	0.41
	E	0.51	0.68	0.03	0.11	0.62	0.50	0.02	0.62	0.12	0.62	0.86	0.17	0.34	0.67	0.05	0.47	0.49	0.05	0.39	0.47	0.45
	O	0.73	0.78	0.39	0.35	0.58	0.68	0.35	0.51	0.53	0.52	0.84	0.44	0.57	0.64	0.36	0.56	0.49	0.21	0.53	0.58	0.58

existence of a context manager for conversation—temporary storage of contextual information for the duration of the conversation. Even if the conversation is relatively short, it may translate to hundreds of tokens of the transcript, which far exceeds the ability of contemporary NLP models to manage the context of the input stream.

Unfortunately, precise fine-tuning of NER models is not a viable option due to the scarcity of annotated spontaneous speech transcripts. There are only a few annotated datasets with NER spans in transcripts of spontaneous speech. The assumption that one will find a similar or close domain is unrealistic. Thus, the choice is difficult: train a NER model from scratch and fine-tune it on out-of-domain datasets, train a NER model from scratch and fine-tune it only on domain data, or use a pre-trained model and fine-tune it with all available data. The results of our experiments do not support the hypothesis that named entities converge across domains of spoken conversations. The way named entities are mentioned in spontaneous speech is domain-dependent. There seems to be no universal patterns appearing in the vicinity of named entity spans across domains, as well as no benefits from pre-training the model on the NER task in a different domain. Fine-tuning with multiple datasets might be beneficial, but this would require coherent normalizations across datasets.

To summarize the subject of NER model fine-tuning for spontaneous conversations, based on the conducted experiments we can formulate the following general guidelines. Using a NER model pre-trained on *OntoNotes v5* without further fine-tuning tends to yield poor results. The most critical aspect of fine-tuning is ensuring it is performed with in-domain data, as adding multiple out-of-domain datasets does not contribute positively. While fine-tuning a model initially trained on *OntoNotes v5* can improve its F_1 score compared to the pure pre-trained version, this improvement mainly addresses issues introduced during the pre-training process and does not significantly enhance the model's ability to generalize. As the result, fine-tuning a model trained from scratch is usually the better approach. Additionally, NER mentions in conversational transcripts often fail to converge, likely due to the inadequacy of available datasets to support better generalization. Finally, Transformer models struggle to overcome discrepancies in transcript normalization.

The last finding is probably the most controversial. As is commonly known, the normalization of input data is an essential pre-processing step for training language models. The purpose of normalization is to standardize the input data and remove any noise or variations that may affect the model's performance. Normalization techniques include lower-casing the text, removing punctuation, and replacing contractions with their expanded forms. Additionally, numerical values and dates may be replaced with generic tokens to reduce the vocabulary size. It is reasonable to assume that the training data for `RoBERTa` should have included various ways in which cardinals, dates, intervals, and times could be written. Yet, one observes a very large variability in the performance of Transformer-based models for entity classes which may have different normalizations.

Results presented in this section have certain limitations. The experiments are conducted on a single baseline model. `RoBERTa` is a popular choice for NER models in the NLP community, but without more extensive experimentation, it cannot be assumed that the findings hold for other model architectures. Obviously, the presented results are relevant only for English. Other languages, particularly languages with more complex syntax and grammar, may allow for the generalization of named entities between datasets. English lacks conjugation, declination, and grammatical genders, and these linguistic features may be helpful in entity recognition in other languages. The reader should also remember that all experiments, results, and conclusions are valid for spontaneous human conversations only and should not be applied to the domain of voice assistants. The structure of utterances when interacting with voice assistance (or any other non-human voice-based interface) is very different from spontaneous speech. These utterances are short, non-conversational phrases, most often expressed in the imperative mode, with little to no artifacts of spontaneous speech. As such, they definitely present less challenge for NER models. Finally, the experiments are conducted on three datasets with entity spans annotated in spontaneous speech transcripts. Although these are primary datasets to train NER models for spontaneous speech, the conclusions presented above may be invalidated as more annotated datasets become available.

6.3 Artifacts of ASR Errors in NER

As we have previously shown, transcripts of spontaneous human speech present a significant obstacle for traditional NER models. In this section, we present the complex relationship between ASR and NER errors which limit the ability of NER models to recover entity mentions from spontaneous speech transcripts. Using publicly available benchmark datasets (SWNE, Earnings-21, OntoNotes), we introduce the taxonomy of ASR-NER errors and measure their true impact on entity recognition. The main finding of presented research is that NER models fail to recognize entity spans even if no word errors are introduced by the ASR. As the consequence, the idea of using the F_1 score to evaluate NER models on conversational transcripts becomes very questionable. Research presented in this section was led by Szymański [41].

We have seen several times that the performance of NLP models tends to deteriorate significantly when the models are applied to the raw outputs of the ASR system. This is particularly true for NER models, which struggle to retrieve even the most basic entity mentions from spontaneous speech. There are three primary factors that contribute to the existence of the gap between ASR output and NLP models. The first factor has been already mentioned in this book several times, it is the fundamentally different grammatical structure of the spoken and spontaneous language when compared to the prescriptive written language. The second challenge comes from the original ASR output containing neither punctuation nor sentence segmentation. These have to be restored by an auxiliary downstream model. Thus, NLP models trained on prescriptive written text or scripted conversations already have to process the out-of-domain input (although the most recent end-to-end ASR systems tend to solve this problem by restoring both punctuation and true casing of the transcript). The third problem stems from ASR systems injecting word errors into the transcript. Due to efficiency requirements, most ASR systems use unsophisticated language models such as n-gram models with limited vocabulary. Thus, many utterances in the input audio may be unrecognized and deleted from the output, while other utterances may cause substitutions or insertions of erroneous tokens into the output.

Consider the following sentence: "*I am to see [Dr Smith]*$^{\mathtt{PERSON}}$ *at [9 am]*$^{\mathtt{TIME}}$ *on [Monday, May 14th]*$^{\mathtt{DATE}}$". The NER model[2] correctly recognizes three entity spans in the sentence. Compare this to the NER spans recognized in the sentence, which is far more likely to be produced by the ASR: "*I am to see doctor [Smith]*$^{\mathtt{PERSON}}$ *at nine I am on [monday]*$^{\mathtt{DATE}}$ [uhm]$^{\mathtt{ORG}}$ yeah [monday]$^{\mathtt{DATE}}$ *may for teen.*" Two entity spans have been cut short, an incorrect label has replaced one span's label, and the model recognized a filler *uhm* as the entity `ORG`. With a few more ASR errors and lowercase output, the model does not recognize a single entity in the output of the ASR: "*I am to see doctor uhm doctor smith at nine I am on man day may for teen.*"

The main problem is that ASR errors are very "unnatural" from the point of view of the NER model because they tend to break the grammar of the sentence on which the NER model depends. One of the most consequential errors made by the ASR is the confusion about the part-of-speech tag. Consider possible ASR errors in the sentence "*My [second]*$^{\mathtt{ORDINAL}}$ *visit is [Wednesday]*$^{\mathtt{DATE}}$ *at [half past one]*$^{\mathtt{TIME}}$." Changing the personal pronoun "*My*" to the noun "*May*" forces the NER model to recognize a `DATE` span, which is reasonable. But if the ASR changes the preposition "*at*" into a verb "*add*," the NER model loses the ability to recognize the utterance "*half past one*" as `TIME` because of the lack of the preceding preposition. Similarly, changing "*half past one*" to "*[one thirty]*$^{\mathtt{TIME}}$" retrieves the `TIME` span, but an ASR error confusing the numeral "*one*" with the conjunction "*when*" produces "*[Wednesday]*$^{\mathtt{DATE}}$ *at when [thirty]*$^{\mathtt{DATE}}$." If, however, the same word is mistakenly recognized as the verb "*want,*" the NER model produces "*[Wednesday]*$^{\mathtt{DATE}}$ *at want [thirty]*$^{\mathtt{CARDINAL}}$".

Unfortunately, the problems mentioned above cannot be easily solved. Word error rates of ASR systems remain high for spontaneous human conversations [11]. The interplay of ASR errors and spontaneous speech characteristics makes the processing of spontaneous speech transcripts with NLP models so challenging. On the one hand, every NLP model is inherently flawed and produces errors (such as not recognizing an instance of an entity). On the other hand, the ASR system injects errors in the form of insertions, deletions, and substitutions. This changes the structure and semantics of transcribed speech and introduces yet another source of errors: alignment. In order to measure the quality of the NER model on the transcript, one has to align tokens between gold transcripts and the ASR output to match entity spans. This process may produce artifacts that significantly skew the results of the evaluation.

The evaluation of the NER task is usually performed using precision, recall, and the F_1 score. Unfortunately, these measures are of limited use for processing spontaneous conversation transcripts because they confound two independent factors contributing to the errors mentioned above: the inability of the NER model to recognize a span as an entity and the word error introduced by the wrong transcription

---

[2] In this illustration we are using `spaCy` [45] trained on OntoNotes v5, Wordnet 3.0, and ClearNLP Constituent-to-Dependency Conversion [8].

of a token. The artifacts of ASR errors and misalignment demonstrate themselves clearly in the NER task and influence the way NER models should be evaluated for spontaneous transcripts. Let us discuss the problem of entity alignment first.

The evaluation of NER models is measured by the loss of entity spans recognized in the ASR output compared to those recognized in the gold transcript. Thus, tokens must be aligned between the ASR output and the gold transcript, as the two may differ in the number of tokens. Alignment is performed after diarization (separating speakers' utterances into separate channels) for each channel independently. Usually, the greedy alignment procedure is used to solve this task. First, the NER model is applied to the gold transcript, producing a tag for each token in the transcript using the `BIO` scheme (`B`—beginning of an entity span, `I`—inside an entity span, `O`—outside of an entity span). Next, all adjacent `I`-tags are collapsed so that each channel is represented by a sequence of `B`-tags and `O`-tags. The same procedure is repeated for the ASR output. The alignment of gold transcripts, normalized gold transcripts, and the ASR output is performed by the `fstalign` [24] and the `kaldialign` [55] libraries, with minor additional corrections. All transcripts are matched at the level of tokens.

6.3.1 Types of ASR Errors

What types of ASR errors can be introduced into the transcript? Following the terminology of Pallett [29], we will distinguish between the following errors:

- *insertion*: a token has been inserted into the ASR output which does not appear in the gold transcript,
- *substitution*: a token has been wrongly transcribed, the number of tokens in both transcripts is the same, but the values of tokens differ,
- *deletion*: the ASR has not recognized a token, the output sequence of the ASR is shorter than the original gold transcript.

At the same time, the NER model can independently introduce the following errors:

- *hallucination*: an entity tag has been produced in the ASR output which does not appear in the gold transcript,
- *replacement*: an entity tag has been added to the token, but the label of the entity class is different from the gold transcript,
- *omission*: the NER model does not produce an entity tag for a token tagged in the gold transcript.

Let us now describe in detail all possible combinations of the above ASR and NLP errors and their impact on the recognition of named entities. Let us consider a scenario where the gold transcript and the ASR output are perfectly aligned, i.e., all tokens are correctly recognized. The gold transcript contains the utterance

Table 6.10 NER errors for fully aligned transcripts (source: [41])

	Second	Quarter	Twenty	Twenty
Gold transcript	B-DATE	I-DATE	I-DATE	I-DATE
Fully matched	B-DATE	I-DATE	I-DATE	I-DATE
Fully omitted	O	O	O	O
Fully replaced	B-CARD	I-CARD	I-CARD	I-CARD
Partially replaced	B-DATE	I-DATE	B-CARD	I-CARD
Partially omitted	B-DATE	I-DATE	O	O
Partially replaced and omitted	B-DATE	I-DATE	O	B-CARD

"*second*$^{\texttt{B-DATE}}$ *quarter*$^{\texttt{B-DATE}}$ *twenty*$^{\texttt{B-DATE}}$ *twenty*$^{\texttt{B-DATE}}$." The following entity span errors are possible (Table 6.10):

- *full match*: each token in the ASR output receives the same entity tag as the gold transcript,
- *full omission*: no entity tags are produced for tokens inside the gold transcript entity span,
- *full replacement*: each token in the ASR output has a different entity tag from the gold transcript,
- *partial match with replacement*: some tokens in the ASR output have different entity tags from the gold transcript,
- *partial match with omission*: some tokens in the ASR output do not have entity tags,
- *partial match with omission and replacement*: some tokens in the ASR output have a different entity class tag, and some tokens do not have entity tags.

Consider a situation where the ASR inserts a token into the gold transcript. Obviously, there is a mismatch in the number of tokens in the gold transcript and the transcription. Let us assume that the utterance "*nextstart*$^{\texttt{B-ORG}}$ *group*$^{\texttt{I-ORG}}$" has been mistakenly transcribed as "*next door group.*" Table 6.11 summarizes possible combinations of ASR and NER errors.

- *full match*: tokens are tagged with the same entity class labels,
- *full omission*: the introduction of a token by the ASR prevents the NER model from finding any entity tags,
- *full substitution*: tag introduced by the ASR forces the NER model to generate different entity labels,
- *partial substitution*: some tokens in the ASR output are tagged with different entity class labels,
- *partial omission*: some tokens in the ASR output do not have an entity tag, which may result in the multiplication of the entity span or shortening of the entity span.

Table 6.11 NER errors for transcripts with ASR insertion (source: [41])

	Nextstart next	Door	Group group
Gold transcript	B-ORG		ORG
Fully matched	B-ORG	I-ORG	I-ORG
Fully omitted	O	O	O
Fully substituted	B-PROD	I-PROD	I-PROD
Partially substituted	B-ORG	I-ORG	B-LOC
Partially omitted	B-ORG	O	B-ORG
Partially omitted	B-ORG	O	O

Table 6.12 NER errors for transcripts with ASR deletion (source: [41])

	American	Door	Bell	Group
	American	<del>	Bell	Group
Gold transcript	B-ORG	I-ORG	I-ORG	I-ORG
Partially matched	B-ORG		I-ORG	I-ORG
Fully omitted	O		O	O
Partially replaced	B-GPE		B-ORG	I-ORG
Partially matched with omission	B-ORG		I-ORG	O
Partially matched with replacement and omission	B-GPE		O	B-ORG

The ASR can delete a token from the gold transcript, resulting in a possible misalignment. In this scenario, full matching is impossible because the gold transcript will contain an unmatched token. Similarly, an entity span cannot be hallucinated or fully substituted. Let us assume that the gold transcript utterance "*american*$^{\texttt{B-ORG}}$ *door*$^{\texttt{I-ORG}}$ *bell*$^{\texttt{I-ORG}}$ *group*$^{\texttt{I-ORG}}$" has been mistakenly transcribed as "*american* < *del* > *bell group*" (i.e., the ASR failed to recognize the "*door*" token). Table 6.12 presents possible combinations of ASR and NER errors.

- *partial match*: tokens not deleted by the ASR have correct entity tags,
- *full omission*: the deletion of a token by the ASR prevents the NER model from producing any entity tags,
- *partial replacement*: some tokens in the ASR output have the wrong entity tag,
- *partial omission*: the loss of token results in some of the tokens not being tagged with an entity tag,
- *partial replacement and omission*: some of the tokens receive correct entity tags, some receive wrong entity tags, and some do not receive any entity tags at all.

The number of possible mistakes is large, and it is not obvious which scenarios are common or rare. In other words, if one is to develop more robust models for named entity recognition in the transcripts of spontaneous speech, it is necessary to understand which scenarios are the most impactful for the NER task. Readers interested in analyzing real-world examples of ASR-NER errors can refer to Appendix A where we present multiple instances of such errors found in the popular *Earnings-21* dataset.

Let us now examine the results of the experimental evaluation of the impact of different ASR errors on the performance of NER. The experiments reported in this section are conducted on the same datasets as the experiments described in Sect. 6.2.1. They are: *OntoNotes*, *SWNE* (Switchboard Dialog Act Corpus annotated with named entity tags), and *Earnings-21*. Again, we omit *CoNLL-2003/CoNLL++* dataset because it is annotated with only four entity types. One may argue that these datasets are not representative of spontaneous conversations. For instance, *Earnings-21* transcripts sound heavily scripted, and the interlocutors present speeches rather than a free exchange of utterances. While this is true, at the same time, these three datasets present the closest that researchers can get to conversational audio transcripts with annotated entity spans.

There are datasets with audio recordings annotated with entity spans, but these datasets are not in the domain of spontaneous speech. In recent years we are observing significant progress in named entity recognition in transcripts of scripted speech. This progress is made possible mostly due to the publication of annotated datasets. Yadav *et al.* [51] present a dataset consisting of TED talks, Mozilla Common Voice recordings, LibriSpeech audiobook recordings, and VoxForge recordings. As the authors observe, NER models achieve promising results on these transcripts (probably due to the fact that the input transcript is semantically similar to the typical training data for NER models). The same dataset is used by Zhang *et al.* [54] to illustrate the error correction model. Recently, annotated transcripts of speech (albeit non-conversional) have been released for Scandinavian languages [32], for French [25], and for Chinese [6]. It is worth mentioning that NER task has been added to the recent Spoken Language Understanding Evaluation (SLUE) benchmark [36]. Unfortunately, the annotation covers a small subset of the *VoxPopuli* dataset, which is not representative of spontaneous speech, the *VoxPopuli* is the set of recorded speeches in the European Parliament.

Entity classes annotated in the above datasets can be broadly divided into closed-domain and open-domain types. Closed-domain entity classes can be regarded as almost gazetteers, i.e., these are classes for which a vast majority of entities can be listed. Examples of closed-domain entity classes include geographical locations or first names (since the distribution of US first names follows a power law distribution [16], a relatively small number of first names represents the majority of first names encountered in the dataset). On the other hand, open-domain entity classes cannot be summarized using a gazetteer. This is the case with numbers, product names, money, or organizations. Unfortunately, gazetteers are not a viable solution even for closed-domain entity classes because ASR errors may produce tokens outside the gazetteer. One possible solution would be to try to overcome ASR errors by retrofitting token representations using domain datasets. This technique, described in Sect. 5.4, has been successfully applied to static word embeddings to mitigate ASR errors.

One might argue that the most important variable influencing the performance of downstream NLP tasks on a transcript is the choice of a particular ASR system. However, this does not seem to be the case. The ASR-NLP gap is equally pronounced for all major commercial ASR systems. The results presented in this section were computed using five different state-of-the-art commercial ASR solutions, selecting telephony-related models if possible. Due to licensing restrictions, which prevent scientists from comparing most of commercial ASR offerings on non-public data sets, the names of particular ASRs could not be disclosed. In order to present any results, one of the commercial ASRs had to be chosen without comparing it to other solutions. In the remainder of the section the results will be reported for the ASR offered Microsoft due to its lowest reported WER on the *Earnings-21* dataset.

6.3.2 Experiments on Gold Transcripts

The first experiment evaluates the state-of-the-art NER model on gold transcripts. A Transformer is trained using the `roberta-large` architecture [22] on the train split of the *OntoNotes* dataset.[3] The evaluation is performed on *Earnings-21*, *SWNE*, and the test split of the *OntoNotes* datasets (see Table 6.13). In order to make the comparison as fair as possible, gold transcripts are normalized using a set of heuristics. Normalization changes all numbers into respective words. The positions of the currency indicator are unified when spelling monetary values, the same normalization is applied to the position of the percent sign. All gold transcripts are properly cased and punctuated. The results are reported as measured by the micro F_1 score because the dataset is highly imbalanced. These experimental settings are very favorable for the ASR. Not only is the transcript fully normalized, but the alignment procedure is fine-tuned to reduce the number of misalignments as much as possible. Furthermore, the NER model is applied to text fragments chunked according to punctuation in the gold transcripts and not to fixed-width sliding windows. In other words, the NER model is applied to the input text of much higher quality than should be expected from the commercial ASR.

Despite the fact that *OntoNotes* contains a significant amount of transcripts of unscripted human conversations, the accuracy of the model deteriorates dramatically on *SWNE* and *Earnings-21* datasets. For all entity classes, the recognition in *SWNE* and *Earnings-21* is much lower than for the *OntoNotes*. The NER model struggles particularly with open-domain entity classes. The complete failure to recognize `MONEY`, `PRODUCT` or `TIME` entities makes the NER model practically unusable in real-world scenarios. Leaving aside more exotic classes represented in the data by a few examples (`LANGUAGE`, `LAW`, `WORK_OF_ART`), we see that the NER model performs better (albeit not satisfactorily) for closed-domain classes, where it can, to a certain degree, memorize most of the instances of a class. For open-domain entity

[3] Other models, including BERT, DistilBERT, FLERT, and spaCy, were tried, due to space limitations only the best-performing model is used for the presentation of results.

Table 6.13 F_1 scores of the NER model on gold transcripts (source: [41])

Entity	Earnings-21	SWNE	OntoNotes
CARDINAL	0.46	0.69	0.86
DATE	0.49	0.34	0.87
EVENT	0.12	0.37	0.74
FAC	0.07	0.32	0.77
GPE	0.63	0.87	0.97
LANGUAGE	0.00	0.94	0.75
LAW	0.02	0.36	0.67
LOC	0.56	0.45	0.76
MONEY	0.20	0.62	0.90
ORDINAL	0.79	0.00	0.86
ORG	0.49	0.62	0.92
PERCENT	0.66	0.00	0.86
PERSON	0.55	0.82	0.96
PRODUCT	0.10	0.58	0.79
QUANTITY	0.42	0.59	0.79
TIME	0.32	0.39	0.69
WORK_OF_ART	0.00	0.46	0.72
micro avg F_1	0.37	0.51	0.83

classes, the performance of the model is disappointingly bad. Please note that the NER model is applied to properly cased and punctuated transcripts of conversations and not to the ASR output, yet the F_1 scores are significantly lower than the scores obtained on the test split of the *OntoNotes* dataset.

6.3.3 Experiments on ASR Output

The second experiment applies the NER model to the *Earnings-21* dataset and measures the number of occurrences of every NER span alignment error. The results are presented in Table 6.14. The first column reports the number of occurrences of NER model errors when the ASR output is fully matched with the gold transcript (no ASR errors in the transcript). Subsequent columns report the number of occurrences of NER model errors when the ASR output is misaligned with the gold transcript due to token insertion, substitution, or deletion by the ASR. Please note that ASR insertion, substitution, and deletion errors often co-occur within a single entity span in the gold transcript, so a single entity span may contribute to multiple cells in the table. The intention is to show the real impact of each type of ASR-NLP error.

The results presented in Table 6.14 clearly show the importance of the joint ASR-NLP model evaluation, as reflected by the breakdown of the two error sources. First,

Table 6.14 Counts of different combinations of NER-ASR errors on the *Earnings-21* dataset (source: [41])

	No ASR error	ASR insertion	ASR substitution	ASR deletion
Correct tags	11408	64	1008	0
Hallucinated	1039	784	958	200
Omitted	3607	47	2649	709
Replaced	1383	6	509	0
With replacement without omission	97	2	9	0
Without replacement with omission	654	37	261	306
With replacement and omission	26	3	19	18

the NER model makes mistakes on fully matched transcripts of spoken conversations, i.e., when the ASR manages to retrieve the gold transcript in the entity span without errors. These errors are responsible for approximately half of all recorded errors. Let us stress this result again: NER models are inherently incapable of processing the transcripts of spontaneous speech; even if the ASR introduces no errors, 37% of entity spans are partially or fully wrong (first column in Table 6.14). The NER model is also very sensitive to errors introduced by the ASR. It can correctly recognize only 18% of entities when the ASR substitutes a token inside the entity span, 6.8% of entities when the ASR inserts a token inside the entity span, and it fails to correctly recognize an entity when the ASR deletes a token inside the entity span. ASR errors are responsible for many hallucinated entities and the majority of omissions. In practice, the number of entity errors doubles compared to the number of errors made on fully matched transcript: ca. 6200 omitted entities in total versus 3600 with perfect transcript and ca. 2000 hallucinated ones versus 1000 with the perfect transcript. Again, let us reiterate this finding: the NER model is helpless when ASR errors are introduced inside entity spans and cannot retrieve an entity when tokens are inserted, substituted, or deleted from entity spans. These results are vastly different from what one could infer from a WER of 15.8 and entity WER of 20.0 reported in [11]. Finally, partial matches, while less pronounced than hallucinations, replacements, and omissions, are important. The true effect of entity hallucinations and omissions in a joint ASR-NLP system can only be measured on a downstream task. Usually, named entity recognition is a single step in a wider NLP task. This task may have a separate evaluation scheme with different metrics and business objectives. For example, in the task of intent retrieval and slot filling, hallucinating or omitting an entity span can lead to a situation where the intent is either not matched or matched in the wrong place. However, the effect of partial matches is more difficult to evaluate. With partial matching, the intent is caught, and the slot is filled, but most probably, the slot is filled with incorrect values. The scale of failures and the impact of upstream model improvements can only be measured by evaluating the entire NLP pipeline on a reference dataset with annotations of intents and slots. This observation strengthens the belief that measuring the increase in the scale of errors in a joint ASR-NLP system is more important than focusing on technical details of measures such as the F_1 score, WER, or ENTWER.

One thing that needs a more thorough discussion is the choice of the F_1 metric. This metric is questionable in the context of NER recognition in ASR transcripts because it is highly dependent on two factors: the word error rate (WER) produced by the ASR and the definition of span alignment. Consider a gold transcript annotation "*John*$^{\texttt{B-PERSON}}$ *F.*$^{\texttt{I-PERSON}}$ *Kennedy*$^{\texttt{I-PERSON}}$" and the ASR output with "*F.*" transcribed as "*eh*" annotated as follows: "*John*$^{\texttt{B-PERSON}}$ *eh Kennedy*$^{\texttt{B-PERSON}}$." Should this annotation be considered correct? The original person entity starting at "*John*" is only partially matched, and a new person entity starting at "*Kennedy*" is introduced in the ASR output. Consider another gold annotation of the following transcript: "*second*$^{\texttt{B-DATE}}$ *quarter*$^{\texttt{I-DATE}}$ *twenty*$^{\texttt{I-DATE}}$ *twenty*$^{\texttt{I-DATE}}$," which the NER model tags as follows: "*second*$^{\texttt{B-DATE}}$ *quarter*$^{\texttt{I-DATE}}$ *twenty*$^{\texttt{B-CARDINAL}}$ *twenty*$^{\texttt{I-CARDINAL}}$" (NER model trained on written language does not recognize "*twenty twenty*" as a valid date). Again, how should this scenario be scored by an accuracy metric? Unfortunately, the traditional definition of the F_1 score is too restrictive to produce a robust score that could paint a reliable picture of the model's performance. The design and implementation of a metric that could compute the alignment of entity spans in the presence of ASR errors would be a significant step in the direction of producing more robust NER models for spoken conversations.

The results presented in this chapter provide a thorough, albeit pessimistic, reality check on the named entity recognition in conversational transcripts. The first conclusion is straightforward: currently available NER models are not trained on representative data (due to the lack of annotated datasets), and their performance on transcripts of spontaneous conversations is much worse than their performance on written language. Importantly, this failure cannot be attributed solely to the presence of ASR word errors. NER models exhibit very high ENTWERs even on gold transcripts, where no ASR errors are present. When the transcript contains ASR insertions, substitutions, or deletions, the entity recognition rates fall to the level where NER models become unusable in downstream tasks. Secondly, a completely new approach is required to meaningfully measure the quality of NER models on conversational transcripts. Traditional metrics, such as F_1 score or ENTWER do not account for the interplay of factors (NER errors, ASR errors, artifacts of spontaneous speech) and do not provide a useful insight into the model's performance. We need to design a more complex evaluation scheme that would take into account the token alignment errors, partial entity span matchings, ASR word errors, and NER errors.

The limitations of the results presented in this section echo the remarks from Sect. 6.2.1. Obviously, the results apply to transcripts of spontaneous conversations in English only. Since we are investigating the problem of named entity recognition, we have to point out that there are practically no datasets of human conversations (both audio and transcripts) annotated with entity spans apart from *SWNE*, *OntoNotes* and *Earnings-21*, the three datasets used in this section. These datasets are relatively small, and the distribution of the frequency of appearance of entity classes is extremely skewed, with several entity classes represented by a handful of examples.

The experiments are conducted with the ASR on audio files from the *Earnings-21* dataset. These files are recorded at 11 kHz–44 kHz, while typical call center conversations are recorded at 8 kHz–16 kHz. Unfortunately, training datasets with recording characteristics resembling real-world usage scenarios are unavailable. Another problem not addressed in these experiments is the problem of racial, gender, and age disparity [20]. The lack of availability of sufficiently representative and inclusive datasets is a pressing issue. But even without the access to more diversified recordings it is reasonable to expect that the performance of the ASR deteriorates for the recordings of speakers other than male speakers of General American dialect of English.

The final remark concerns a broader ethical issue. The work on spoken language understanding aims at broadening the accessibility of communication technology. Spontaneous spoken language is the least limiting mode of interacting with an information system. This mode does not require any digital competencies or expensive resources. The ability to correctly process spontaneous human conversations opens access to technology to stakeholders who might have been previously excluded. Research community should strive to diminish discrimination resulting from biased training datasets, which may cause specific individuals to be disproportionally mistranscribed due to their accent, dialect, or speech impediments. As digital voice applications become increasingly integrated into society's infrastructure, the need to continuously improve the quality of spoken communications processing models becomes clear. But the ability to better process and understand spoken human conversations carries the significant ethical risk associated with clandestine eavesdropping by adversarial agents. Correct recognition of spoken names of people, places, organizations, or events, can be malevolently used by authoritarian government agencies trying to suppress free speech. Recognition of names of products or services may be utilized by marketers for non-consensual profiling. Thus, it is in the best interest of all to raise public awareness and understanding of computing, the automatic processing of spontaneous speech, and its consequences.

6.4 Closing

It is important to acknowledge the significant impact of recent advancements in generative AI on the field of transcript engineering. Specifically, the emergence of large language models has demonstrated remarkable proficiency in tasks such as transcript segmentation and true casing. These models have shown promising capabilities in addressing more complex tasks like NER, offering partial solutions to this challenge. Another source of progress stems from the development of new end-to-end ASR models. Unlike their predecessors, these models do not produce raw, lowercased, and unpunctuated transcripts. Instead, they use the entire language modeling pipeline to generate outputs of exceptionally high quality.

Naturally, numerous challenges still need resolution, such as ensuring proper diarization and mitigating the influence of background noise, among others. However, the improvements witnessed in the field are encouraging. Let us not forget though that these advancements are predominantly observed for English, while other languages may still lag behind in terms of progress. True, the remarkable AI developments witnessed in recent months have generated substantial expectations surrounding the application of conversational AI. These evolving technologies hold tremendous potential for revolutionizing various domains and augmenting human capabilities.

Lastly, the ethical aspect of transcript engineering warrants careful consideration. Undoubtedly, the advancements in the realm of spoken language understanding hold promising potential but also present new ways for ill-intentioned actors to exploit and harm individuals. Across the globe, authoritarian regimes could employ these technologies to tighten their control over the population, while predatory corporations might use them to exert undue influence on human behavior through AI-driven conversations. The use of large language models brings forth a set of new risks. These models, known to harbor biases and discriminatory tendencies, can exacerbate societal disparities and pose substantial threats, particularly to vulnerable subpopulations [47, 48]. It is increasingly clear among the scientific community that stringent ethical oversight is imperative in this field. Therefore, in order to ensure the responsible and equitable deployment of transcript engineering technologies, it is important to emphasize ethical vigilance throughout all stages of their development and implementation.

References

1. C. Allauzen, M. Riley, J. Schalkwyk, W. Skut, M. Mohri, Openfst: A general and efficient weighted finite-state transducer library: (extended abstract of an invited talk), in *Implementation and Application of Automata: 12th International Conference, CIAA 2007, Praque, Czech Republic, July 16-18, 2007, Revised Selected Papers 12* (Springer, 2007), pp. 11–23
2. R. Ardila, M. Branson, K. Davis, M. Henretty, M. Kohler, J. Meyer, R. Morais, L. Saunders, F.M. Tyers, G. Weber, Common voice: A massively-multilingual speech corpus. *arXiv preprint* arXiv:1912.06670 (2019)
3. S.B. David, Ner evaluation (GitHub repository, 2020)
4. F. Béchet, A.L. Gorin, J.H. Wright, D. Hakkani-Tür, Named entity extraction from spontaneous speech in how may i help you? In *INTERSPEECH* (2002)
5. A. Caubrière, S. Rosset, Y. Estève, A. Laurent, E. Morin, Where are we in named entity recognition from speech? in *Proceedings of the Twelfth Language Resources and Evaluation Conference* (2020), pp. 4514–4520
6. B. Chen, G. Xu, X. Wang, P. Xie, M. Zhang, F. Huang, Aishell-ner: Named entity recognition from chinese speech, in *ICASSP 2022-2022 IEEE International Conference on Acoustics, Speech and Signal Processing (ICASSP)* (IEEE, 2022), pp. 8352–8356
7. J.D. Choi, Switchboard named entity corpus. https://github.com/emorynlp/swne (2020)
8. J.D. Choi, H. Chen, T. Jurczyk, Constituent to dependency conversion. https://github.com/clir/clearnlp-guidelines (2016)
9. C. Cieri, D. Miller, K. Walker, The Fisher corpus: a resource for the next generations of speech-to-text. Int. Conf. Lang. Res. Eval. **4**, 69–71 (2004)

10. X. Dai, Recognizing complex entity mentions: A review and future directions, in *Proceedings of ACL 2018, Student Research Workshop* (2018), pp. 37–44
11. M. Del Rio, N. Delworth, R. Westerman, M. Huang, N. Bhandari, J. Palakapilly, Q. McNamara, J. Dong, P. Zelasko, M. Jetté, Earnings-21: A practical benchmark for asr in the wild. *arXiv preprint* arXiv:2104.11348 (2021)
12. B. Favre, F. Béchet, P. Nocéra, Robust named entity extraction from large spoken archives, in *Proceedings of Human Language Technology Conference and Conference on Empirical Methods in Natural Language Processing* (2005), pp. 491–498
13. S. Ghannay, A. Caubrière, Y. Estève, N. Camelin, E. Simonnet, A. Laurent, E. Morin, End-to-end named entity and semantic concept extraction from speech, in *2018 IEEE Spoken Language Technology Workshop (SLT)* (IEEE, 2018), pp. 692–699
14. A. Goyal, V. Gupta, M. Kumar, Recent named entity recognition and classification techniques: a systematic review. Comput. Sci. Rev. **29**, 21–43 (2018)
15. R. Grishman, B.M. Sundheim, Design of the muc-6 evaluation, in *TIPSTER TEXT PROGRAM PHASE II: Proceedings of a Workshop held at Vienna, Virginia, May 6-8, 1996* (1996), pp. 413–422
16. M.W. Hahn, R. Alexander Bentley, Drift as a mechanism for cultural change: an example from baby names. *Proceedings of the Royal Society of London. Series B: Biological Sciences* (2003), 270(suppl_1):S120–S123
17. M. Hatmi, C. Jacquin, E. Morin, S. Meigner, Incorporating named entity recognition into the speech transcription process, in *Proceedings of the 14th Annual Conference of the International Speech Communication Association (Interspeech'13)* (2013), pp. 3732–3736
18. Z. Ju, J. Wang, F. Zhu, Named entity recognition from biomedical text using svm, in *2011 5th international conference on bioinformatics and biomedical engineering* (IEEE, 2011), pp. 1–4
19. J. Kahn, O. Galibert, L. Quintard, M. Carré, A. Giraudel, P. Joly, A presentation of the repere challenge, in *2012 10th International Workshop on Content-Based Multimedia Indexing (CBMI)* (IEEE, 2012), pp. 1–6
20. A. Koenecke, A. Nam, E. Lake, J. Nudell, M. Quartey, Z. Mengesha, C. Toups, J.R. Rickford, D. Jurafsky, S. Goel, Racial disparities in automated speech recognition. Proceed. Nat. Acad. Sci. **117**(14), 7684–7689 (2020)
21. J. Li, A. Sun, J. Han, C. Li, A survey on deep learning for named entity recognition. IEEE Trans. Knowl. Data Eng. **34**(1), 50–70 (2020)
22. Y. Liu, M. Ott, N. Goyal, J. Du, M. Joshi, D. Chen, O. Levy, M. Lewis, L. Zettlemoyer, V. Stoyanov, Roberta: A robustly optimized bert pretraining approach. *arXiv preprint* arXiv:1907.11692 (2019)
23. M. Marrero, J. Urbano, S. Sánchez-Cuadrado, J. Morato, J.M. Gómez-Berbís, Named entity recognition: fallacies, challenges and opportunities. Comput. Stand. Interf. **35**(5), 482–489 (2013)
24. Q. McNamara, D. Kokotov, fstalign. https://github.com/revdotcom/fstalign (2021)
25. A. Millour, Y. Dupont, A. Jouglar, K. Fort, Fenec: un corpus équilibré pour l'évaluation des entités nommées en français (fenec: a balanced sample corpus for french named entity recognition), in *Actes de la 29e Conférence sur le Traitement Automatique des Langues Naturelles. Volume 1: conférence principale* (2022), pp. 82–94
26. J. Mizgajski, A. Szymczak, R. Głowski, P. Szymański, P. Żelasko, Ł. Augustyniak, M. Morzy, Y. Carmiel, J. Hodson, Ł. Wójciak, et al., Avaya conversational intelligence: A real-time system for spoken language understanding in human-human call center conversations. *arXiv preprint* arXiv:1909.02851 (2019)
27. S. Morwal, N. Jahan, D. Chopra, Named entity recognition using hidden markov model (hmm). Int. J. Nat. Lang. Comput. (IJNLC) **1** (2012)
28. D. Nadeau, S. Sekine, A survey of named entity recognition and classification. Lingvisticae Investigationes **30**(1), 3–26 (2007)
29. D.S. Pallett, Performance assessment of automatic speech recognizers. J. Res. Nat. Bureau Stand. **90**(5), 371 (1985)

30. V. Panayotov, G. Chen, D. Povey, S. Khudanpur, Librispeech: an asr corpus based on public domain audio books, in *2015 IEEE International Conference on Acoustics, Speech and Signal Processing (ICASSP)* (IEEE, 2015), pp. 5206–5210
31. C. Parada, M. Dredze, F. Jelinek, Oov sensitive named-entity recognition in speech, in *Twelfth Annual Conference of the International Speech Communication Association* (2011)
32. D. Porjazovski, J. Leinonen, M. Kurimo, Attention-based end-to-end named entity recognition from speech, in *International Conference on Text, Speech, and Dialogue* (Springer, 2021), pp. 469–480
33. C. Janarish Saju, A.S. Shaja, A survey on efficient extraction of named entities from new domains using big data analytics, in *2017 Second International Conference on Recent Trends and Challenges in Computational Models (ICRTCCM)* (IEEE, 2017), pp. 170–175
34. E.F. Sang, F.D. Meulder, Introduction to the conll-2003 shared task: Language-independent named entity recognition. *arXiv preprint* arXiv:cs/0306050 (2003)
35. D. Serdyuk, Y. Wang, C. Fuegen, A. Kumar, B. Liu, Y. Bengio, Towards end-to-end spoken language understanding, in *2018 IEEE International Conference on Acoustics, Speech and Signal Processing (ICASSP)* (IEEE, 2018), pp. 5754–5758
36. S. Shon, A. Pasad, F. Wu, P. Brusco, Y. Artzi, K. Livescu, K.J. Han, Slue: New benchmark tasks for spoken language understanding evaluation on natural speech, in *ICASSP 2022-2022 IEEE International Conference on Acoustics, Speech and Signal Processing (ICASSP)* (IEEE, 2022), pp. 7927–7931
37. E. Shriberg, Spontaneous speech: How people really talk and why engineers should care, in *Ninth European Conference on Speech Communication and Technology*. Citeseer (2005)
38. M. Stiefel, N. Thang Vu, Enriching asr lattices with pos tags for dependency parsing, in *Proceedings of the Workshop on Speech-Centric Natural Language Processing* (2017), pp. 37–47
39. D. Sui, Z. Tian, Y. Chen, K. Liu, J. Zhao, A large-scale chinese multimodal ner dataset with speech clues, in *Proceedings of the 59th Annual Meeting of the Association for Computational Linguistics and the 11th International Joint Conference on Natural Language Processing (Volume 1: Long Papers)* (2021), pp. 2807–2818
40. M. Surdeanu, J. Turmo, E. Comelles, Named entity recognition from spontaneous open-domain speech, in *INTERSPEECH* (2005), pp. 3433–3436
41. P. Szymański, L. Augustyniak, M. Morzy, A. Szymczak, K. Surdyk, P. Żelasko, Why aren't we ner yet? artifacts of asr errors in named entity recognition in spontaneous speech transcripts, in *Proceedings of the 61st Annual Meeting of the Association for Computational Linguistics (Volume 1: Long Papers)* (2023), pp. 1746–1761
42. P. Szymański, P. Żelasko, M. Morzy, A. Szymczak, M. Żyła-Hoppe, J. Banaszczak, L. Augustyniak, J. Mizgajski, Y. Carmiel, Wer we are and wer we think we are. *arXiv preprint* arXiv:2010.03432 (2020)
43. N. Tomashenko, C. Raymond, A. Caubrière, R. De Mori, Y. Estève, Dialogue history integration into end-to-end signal-to-concept spoken language understanding systems, in *ICASSP 2020-2020 IEEE International Conference on Acoustics, Speech and Signal Processing (ICASSP)* (IEEE, 2020), pp. 8509–8513
44. A. Ushio, J. Camacho-Collados, T-ner: an all-round python library for transformer-based named entity recognition. *arXiv preprint* arXiv:2209.12616 (2022)
45. Y. Vasiliev, *Natural language processing with Python and spaCy: A practical introduction* (No Starch Press, 2020)
46. C. Wang, M. Riviere, A. Lee, A. Wu, C. Talnikar, D. Haziza, M. Williamson, J. Pino, E. Dupoux, Voxpopuli: A large-scale multilingual speech corpus for representation learning, semi-supervised learning and interpretation. *arXiv preprint* arXiv:2101.00390 (2021)
47. L. Weidinger, J. Mellor, M. Rauh, C. Griffin, J. Uesato, P.-S. Huang, M. Cheng, M. Glaese, B. Balle, A. Kasirzadeh, et al., Ethical and social risks of harm from language models. *arXiv preprint* arXiv:2112.04359 (2021)
48. L. Weidinger, J. Uesato, M. Rauh, C. Griffin, P.-S. Huang, J. Mellor, A. Glaese, M. Cheng, B. Balle, A. Kasirzadeh, et al., Taxonomy of risks posed by language models, in *Proceedings of the 2022 ACM Conference on Fairness, Accountability, and Transparency* (2022), pp. 214–229

49. R. Weischedel, M. Palmer, M. Marcus, E. Hovy, S. Pradhan, L. Ramshaw, N. Xue, A. Taylor, J. Kaufman, M. Franchini, et al., Ontonotes release 5.0 ldc2013t19. Linguist. Data Consortium Philadelphia, PA **23**, 170 (2013)
50. Y. Wen, C. Fan, G. Chen, X. Chen, M. Chen, A survey on named entity recognition, in *Communications, Signal Processing, and Systems: Proceedings of the 8th International Conference on Communications, Signal Processing, and Systems 8th* (Springer, 2020), pp. 1803–1810
51. H. Yadav, S. Ghosh, Y. Yu, R. Ratn Shah, End-to-end named entity recognition from english speech. *arXiv preprint* arXiv:2005.11184 (2020)
52. V. Yadav, S. Bethard, A survey on recent advances in named entity recognition from deep learning models. *arXiv preprint* arXiv:1910.11470 (2019)
53. P. Żelasko, J. Mizgajski, M. Morzy, A. Szymczak, P. Szymański, Ł. Augustyniak, Y. Carmiel, Towards better understanding of spontaneous conversations: Overcoming automatic speech recognition errors with intent recognition. *arXiv preprint* arXiv:1908.07888 (2019)
54. F. Zhang, M. Tu, S. Liu, J. Yan, Asr error correction with dual-channel self-supervised learning, in *ICASSP 2022-2022 IEEE International Conference on Acoustics, Speech and Signal Processing (ICASSP)* (IEEE, 2022), pp. 7282–7286
55. P. Żelasko, L. Guo, kaldialign. *GitHub Repository* (2021)

Chapter 7
Conclusions

This book aims to shed light on the domain of spoken language processing, particularly focusing on transcripts of spontaneous conversations. The main goal has been to convey the fundamental distinction between spoken and written language, outlining challenges that spontaneous human conversations pose to the field of NLP, despite recent advancements in this field. By discussing the complexities and nuances of spoken communication, I hope to contribute to the deeper understanding of this multifaceted domain.

Allow me to repeat the primary concept of this book: the significant disparity between spontaneous human speech and written language renders contemporary NLP models, which are primarily trained on written texts, frequently falling short of anticipated performance levels. This discrepancy between oral and written communication is a fundamental characteristic of linguistic systems, and the most viable approach to bridging this divide is the increased investment in linguistic resources tailored for spoken language, including annotated datasets, benchmarks, ontologies, and dictionaries.

The generative AI revolution is still in its early stages, showing tremendous transformative potential. Each day, state-of-the-art benchmarks are being surpassed, signaling a significant shift in the domain of NLP. End-to-end models continue to enhance with each passing month, rendering some aspects discussed in this book outdated. Notably, the Whisper family of models displays remarkable capabilities in true casing and punctuation restoration. Emerging techniques founded on LLM prompting are now rivaling conventional NER models and intent classifiers. Although it is premature to determine the prevailing approach, the ongoing competition presents a fascinating scenario.

This book has primarily concentrated on the English language. It is important to note that advancements in the computational processing of spoken language for lesser-resourced languages lag behind. Apart from major languages such as Chinese or Spanish, the effectiveness of models for the majority of languages is inferior to that

M. Morzy, *Spoken Language Processing*, Studies in Computational Intelligence 1205,
https://doi.org/10.1007/978-3-031-88566-2_7

for English. Even when a specific language task appears to be successfully addressed in English, additional challenges exist for other languages.

The ethical risks and challenges pertaining to the processing of spoken language have not been extensively discussed. With the increasing integration of artificial intelligence across various aspects of daily life, the potential for technology misuse is escalating. Some prominent dangers include fraudulent activities, phishing schemes, unauthorized data access, and illicit surveillance. Intrusive marketing practices also pose significant threats. Conversely, risks may not solely arise from overt threats but can also manifest through discriminatory outcomes. In particular, the performance of AI in handling spoken language may degrade for specific demographic groups, ethnicities, dialects, or individuals with speech impediments. Such deterioration could exacerbate the existing stigmatization experienced by vulnerable populations. The importance of addressing ethical considerations in the realm of technology is a relatively recent development. Observing the efforts to establish ethical guidelines for AI within the domain of spoken language processing promises to be a fascinating attempt.

Verbal communication is a fundamental aspect of human interaction that dates back to the earliest stages of humanity. As we approach the era of technological innovation driven by artificial intelligence, the field of spoken language processing holds immense potential in advancing human welfare. Through leveraging this technology, we can enhance interpersonal communication, extend modern services to individuals with limited digital literacy, offer assistive tools to individuals with disabilities, tailor educational experiences for those with hearing impairments, and enable intuitive interactions with healthcare systems for the elderly. The potential applications are numerous, presenting us with the opportunity to make significant contributions to societal well-being.

As we reach the conclusion of this journey through modern artificial intelligence and spoken language processing, it is hard not to feel both awe and humility. Keeping up with the rapid advancements in AI technology is a truly Sisyphean task, a relentless race that seems to push the limits of our understanding further with each breakthrough. My personal experience from this journey has underscored an essential truth: domain knowledge in language and linguistics, along with a deep understanding of culture and communication, remains paramount. These elements are the bedrock upon which robust and meaningful Conversational AI systems are built. I hold a hopeful vision for the future, where we witness remarkable progress in spoken language processing. A future where the benefits of this transformative technology extend beyond a few privileged languages, reaching communities around the globe. I hope that this technology can build better communication and understanding across diverse cultures and languages. Together, we can look forward to a world where the power of language bridges gaps and brings us closer than ever before.

Appendix A
Examples of ASR-NER Errors from the *Earnings-21* Dataset

Earnings-21 is a popular dataset of spontaneous conversations with entity spans annotated in the transcripts. This section presents several examples of alignments of the ASR output with the gold transcript with entity tags. In each table, the upper two rows present entity tags and word tokens present in the gold transcript, and the bottom two rows present word tokens generated by the ASR and entity tags produced by the NER model. A detailed description of each case is presented in the caption of each table. All examples are taken from [12].

Table A.1 Full matching of word tokens and entity tags

O	O	B-PERSON	O	O	O
Thank	you	anna	and	welcome	everyone
Thank	you	anna	and	welcome	everyone
O	O	B-PERSON	O	O	O

Table A.2 Full matching of entity tags despite the insertion of a token by the ASR

O	B-DATE	I-DATE	I-DATE	I-DATE	B-DATE	O
From	last	<ins>	years	comparable	quarter	results
From	last	year	’s	comparable	quarter	results
O	B-DATE	I-DATE	I-DATE	I-DATE	I-DATE	O

Table A.3 Full matching of entity tags despite the ASR substitution of a token

O	O	B-PERSON	I-PERSON	O	O	O
We	have	dominic	macklon	our	senior	vice
We	have	dominic	macklin	our	senior	vice
O	O	B-PERSON	I-PERSON	O	O	O

M. Morzy, *Spoken Language Processing*, Studies in Computational Intelligence 1205,
https://doi.org/10.1007/978-3-031-88566-2

Table A.4 Full matching of word tokens, the NER hallucinates the `CARDINAL` entity

O	O	O	O	O	O
Your	normal	mid	teens	revenue	growth
Your	normal	mid	teens	revenue	growth
O	O	B-CARDINAL	I-CARDINAL	O	O

Table A.5 The ASR token insertion (due to wrong recognition of "*perishables*" as "*paris rivers*") makes the NER to hallucinate the `GPE` entity

O	O	O	B-ORDINAL
From	<ins>	perishables	first
From	paris	rivers	first
O	B-GPE	O	B-ORDINAL

Table A.6 The ASR deletes a token by recognizing "*good more lean work*" as "*good morning work*", causing the NER to hallucinate the `TIME` entity

O	O	O	O	O	O	O	O
For	the	good	more	lean	work	to	help
For	the	good	<del>	morning	work	to	help
O	O	B-TIME	O	I-TIME	O	O	O

Table A.7 The NER hallucinates the `PERSON` tag due to an ASR substitution

O	O	O	O	O
Now	so	are	there	discernible
Tina	so	are	there	discernible
B-PERSON	O	O	O	O

Table A.8 The `DATE` entity is missed due to the ASR insertion and replacement

O	O	B-ORG	B-DATE	O
<ins>	see	nexstar's	annual	report
Sing	next	cars	annual	report
O	O	O	O	O

Table A.9 The ASR replaces tokens in the unrecognized person's name forcing the NER to omit the `PERSON` entity

B-PERSON	I-PERSON	O	O	O	O
Shuang	liu	and	chief	financial	officer
Strong	will	and	chief	financial	officer
O	O	O	O	O	O

Table A.10 The ASR deletes tokens related to the unrecognized name of the SME company, forcing the NER to omit the `ORG` entity

O	O	O	B-ORG	I-ORG	I-ORG
Profile	to	the	s	m	e
Profile	to	the	s	m	<del>
O	O	O	O	O	O

Table A.11 Full matching of tokens does not prevent the NER from replacing the `DATE` entity with the `CARDINAL` entity

O	B-DATE	I-DATE
In	twenty	nineteen
In	twenty	nineteen
O	B-CARDINAL	I-CARDINAL

Table A.12 The ASR insertion results in the replacement of the `ORG` entity with the `PERSON` entity

O	B-ORG	I-ORG	I-ORG	O	O
And	jj	<ins>	bistricer	chief	operating
And	jj	best	research	chief	operating
O	B-PERSON	I-PERSON	I-PERSON	O	O

Table A.13 The ASR substitution causes the full replacement of the `GPE` entity with the `NORP` entity

O	O	B-GPE	O	O	O
It's	not	mexico	for	example	right
He's	not	mexican	for	example	right
O	O	B-NORP	O	O	O

Table A.14 Example of a partial `DATE` entity match with the rest of the entity replaced by the `CARDINAL` entity despite the full matching of word tokens

B-DATE	I-DATE	I-DATE	I-DATE
Twenty	twenty	second	quarter
Twenty	twenty	second	quarter
B-CARDINAL	I-CARDINAL	B-DATE	B-DATE

Table A.15 Example of partial `CARDINAL` entity match with the replacement of the rest of the entity with `FAC` entity caused by the ASR substitutions

O	B-CARDINAL	I-CARDINAL	I-CARDINAL	O	O
And	one	twenty	eight	total	net
And	waterman	twenty	eight	dot	net
O	B-FAC	B-CARDINAL	I-CARDINAL	O	O

Table A.16 Example of the partial ORG entity match with parts of the entity span omitted despite the full matching of word tokens

O	B-ORG	I-ORG	O	O
while	ingersoll	rand	took	share
while	ingersoll	rand	took	share
O	B-ORG	O	O	O

Table A.17 Example of the partial ORG entity match with parts of the entity span omitted due to ASR insertion and substitution

B-ORG	I-ORG	I-ORG	I-ORG	I-ORG	O
The	<ins>	nextera	energy	inc	and
The	next	era	energy	inc	and
O	O	B-ORG	I-ORG	I-ORG	O

Table A.18 Example of the partial ORG entity match with parts of the entity span omitted or replaced despite the full matching of word tokens

O	O	O	B-ORG	I-ORG	I-ORG	I-ORG	I-ORG
Present	that	to	the	florida	public	service	commission
Present	that	to	the	florida	public	service	commission
O	O	O	O	B-GPE	B-ORG	I-ORG	I-ORG

Table A.19 Example of the partial DATE entity match parts of the entity span omitted or replaced due to ASR substitutions

B-DATE	I-DATE	I-DATE	I-DATE	I-DATE	I-DATE	O
The	second	half	of	twenty	one	operating
The	second	half	i'm	twenty	one	operating
O	O	O	O	B-CARDINAL	B-DATE	O

Table A.20 Example of the partial entity matches with replacements and omissions due to ASR substitutions and deletions

O	B-DATE	I-DATE	I-DATE	I-DATE	B-MONEY	I-MONEY
To	june	30	twenty	twenty	$25.2	million
To	june	3020	twenty	<del>	$25.2	million
O	B-DATE	I-DATE	B-MONEY	O	I-MONEY	I-MONEY

Appendix B
Annotated Datasets for Spoken Language Understanding

In this appendix, we provide an overview of the most important annotated datasets for training conversational models. This list contains various datasets designed for automatic speech recognition as well as datasets with annotations of dialogue acts, named entities, and more. Getting to know and understand datasets is crucial for designing better models. By familiarizing ourselves with the datasets, we can gain a deeper understanding of the problem at hand and the type of data that is available. This can help us identify any potential biases or limitations in the data, which can affect the accuracy and fairness of the model. Understanding the limitations of a dataset can help us design models that are more robust and generalizable.

LibriSpeech

In the domain of ASR, the need for high-quality training data is a challenge that frequently hampers progress. The introduction of the LibriSpeech dataset tries to address this issue. LibriSpeech [8] is a large collection of audio files paired with corresponding transcriptions, totalling 1000 hours of content. This resource originates from LibriVox, a volunteer-driven initiative committed to converting literary works from Project Gutenberg into audiobooks. The creators of the LibriSpeech designed a systematic process to handle text and synchronize audio segments with their corresponding scripts. A distinctive feature of the work includes the development of language models derived from the same literary sources, supporting the methodology with various experiments that demonstrate the effectiveness of the dataset.

Confronted with the problem of inadequate training resources for ASR systems, the authors responded by assembling a corpus rooted in the LibriVox project. LibriVox uses the Project Gutenberg collection, featuring over 8000 free books, to generate audiobooks. The outcome is a dataset of 1000 hours of audio aligned with text, offered under the Creative Commons BY 4.0 license. The audio is sampled at the

M. Morzy, *Spoken Language Processing*, Studies in Computational Intelligence 1205,
https://doi.org/10.1007/978-3-031-88566-2

quality rate of 16 kHz. In their effort to optimize the text for alignment, the authors developed several preprocessing steps. They converted all text to uppercase, stripped away punctuation, and expanded abbreviations as well as rare words. For language modeling, they employed the Witten-Bell bigram model [11], while the Sequitur G2P toolkit handled the pronunciation of out-of-vocabulary words. To facilitate synchronization, audio segments were divided into 30 s portions, and the Smith-Waterman alignment algorithm [10] was utilized to match these segments with text. Further refining the dataset, a secondary filtering stage was enacted to eliminate segments prone to inaccuracy, stemming from potential discrepancies due to errors in Project Gutenberg texts or unintended modifications during reading.

The corpus is structured as follows. Twenty male and twenty female voices were chosen for inclusion in both the development and test sets. Each reader contributed approximately 8 min of audio, resulting in the combined duration of 5 h and 20 min for each set. The rest of the recordings were apportioned into two training sets of 100 and 360 hours each, focusing on "clean" readers, defined by their low WERs. Similarly, a distinct classification was applied to "other" segments, comprising 500 hours of audio by readers with higher WERs. To enhance language modeling, the authors expanded their efforts beyond individual books. They trained models using data from 14500 books in Project Gutenberg, totaling 800 million tokens and containing 900000 unique words. The working vocabulary was restricted to 200000 words, of which one-third had verified pronunciations in the CMU dictionary. Through this methodology, 3-gram and 4-gram language models were developed. The results demonstrated impressive WERs: 5.51% on the test-clean dataset and 13.97% on the test-other dataset, marking a significant achievement in ASR model training efficacy.

CommonVoice

CommonVoice [1] stands out as the most extensive open-source multilingual voice dataset available, aiming to address the significant gap in resources for transcribed audio across multiple languages. The project is dynamic, with new languages being consistently incorporated. The architecture of the CommonVoice dataset reflects the systematic approach utilized to collect and annotate audio samples. The CommonVoice initiative is driven by a community-based effort, using both web and mobile applications to enable a collaborative process. Participants engage in four primary tasks: recording utterances, reviewing them, collecting texts to be read, and reviewing those texts. Each text and recording undergoes a validation process involving up to three contributors who vote on the quality. Clear annotation guidelines are provided to maintain consistency and accuracy. The goal of CommonVoice is to establish an open database with a diverse array of voice samples. It is important to note that the dataset is composed of discrete, non-conversational utterances from single speakers.

The dataset, as of the time of writing, encompasses 28000 hours of recordings across 112 languages, with 18000 hours having been validated. The audio clips are stored as mono-channel, 16-bit MPG3 files, sampled at 48 kHz. The selection of

MPG format, as opposed to lossless compression, is attributed to its widespread compatibility with web browsers, facilitating easier access and usage.

TED-LIUM

TED-LIUM 3 [6] is a large audio dataset curated from the archive of TED Talks. It contains an extensive collection of 2351 audio talks presented in the NIST sphere format (SPH), which builds upon the prior contents of TED-LIUM 2. This compilation totals 452 hours of audio recordings. Accompanying these audio files are aligned automatic transcripts provided in STM format. Additionally, the dataset includes development and test data from TED-LIUM 2, comprising 19 TED talks in SPH format paired with corresponding manual transcriptions. A dictionary containing pronunciations with 159848 entries is also included, consistent with its predecessor, TED-LIUM 2. For language modeling, selected monolingual data has been sourced from publicly available corpora from WMT12, modified for more effective English tokenization. Initially, TED Talk audio and transcripts were released in two phases: first in 2012 with 118 hours of audio and later in 2014 with an additional 207 hours. The release of TED-LIUM 3 marks the final iteration, with new alignments and augmentation of audio, resulting in a repository of 452 hours of speech content.

The challenge addressed by TED-LIUM 3 is the inaccuracy of TED Talk closed captions when utilized as verbatim transcripts. These captions often omit speech disfluencies, such as repetitions and hesitations, and may inaccurately represent expressions or contractions. Being tailored for on-screen reading, the segmentations are not optimized for speech recognition, leading to imprecise timing extractions. To mitigate these issues, the authors aligned 1495 TED Talks using the Kaldi toolkit [9]. This initial alignment was further refined using a model trained on the prior TED-LIUM 1 and 2 datasets, achieving an audio coverage of 83%. The dataset construction emphasizes speaker adaptation, trying to diminish the mismatch between the acoustic model and audio fidelity specific to individual speakers. This balances the dataset in terms of the number of speakers, gender distribution, and the duration of audio from each speaker.

Two models were developed using the original publication of TED-LIUM 3: a 4-gram model with 10 million n-grams and a RNN model with letter-based features, lattice rescoring, and 10 million parameters. The 4-gram model attained a WER of 8% on the test set, while the RNN achieved a WER of 6.9%. Further experimentation involved an end-to-end ASR system reminiscent of Deep Speech 2. This system utilizes input spectrograms, both one-dimensional and two-dimensional convolutions, GRU/LSTM layers (unidirectional and bidirectional), abetted by a lookahead convolution, followed by feed-forward layers and a softmax layer. The end-to-end solution functions independently of a separate language model, registering a WER of 13.7 and a CER of 6.1.

VoxCeleb

The VoxCeleb dataset [7] is an important advancement in speaker recognition, offering a large audio-visual repository sourced from diverse, real-world environments. It tackles the challenges of identifying speakers in unconstrained settings, making it a useful tool for researchers aiming to enhance speech recognition technologies. VoxCeleb comprises more than 7000 speakers, accumulating over 1 million utterances extracted from YouTube interviews. This dataset features a rich variety of accents, languages, ethnicities, and professions, ensuring both accuracy and generalizability. It includes two main versions: VoxCeleb 1 and VoxCeleb 2. VoxCeleb 1 contains over 150000 utterances from 1251 speakers. VoxCeleb 2 expands on this by offering over 1 million utterances from more than 6000 speakers. The VoxCeleb dataset stands out due to its fully automated compilation process, eliminating the need for manual annotation. Initially, potential speakers are identified using the VGGFace datasets, focusing on individuals with a significant online presence. Videos are then retrieved from YouTube using these names combined with keywords like "interview," ensuring relevant content. Face detection algorithms track faces in each video frame. These tracks are verified against known identities using pre-trained VGGFace CNNs.

The SyncNet CNN model is employed to validate audio-visual synchronization by analyzing mouth motions, weeding out non-authentic content. To reduce redundancy, the dataset also removes near-identical speech segments using feature embeddings from a speaker recognition model. In addition, nationality labels are assigned to speakers using publicly available data, enriching the dataset with insights into accent diversity. VoxCeleb features speakers from 145 nationalities and provides a broad linguistic spectrum for model training. The combination of audio and visual data makes it invaluable for audiovisual speech recognition, lip reading, and cross-modal learning. Additional useful feature is that the recordings are caputred in natural, uncontrolled environments with real-world noise, which strengthens the robustness of models trained on it.

Fisher

The Fisher dataset [4] is one of the most important resources in automatic speech recognition. It has been created under the DARPA EARS program to provide a rich training dataset for near-spontaneous conversation analysis. Comprising over 16000 conversations and more than 2000 hours of transcribed speech, this dataset is distinguished by its diversity in gender, age, language, and conversation topics, covering 100 distinct themes. The Fisher dataset was designed based on the Fisher collection protocol, which involved numerous participants engaging in short calls with strangers on pre-defined topics, changing every 24 hours. The DARPA EARS program aimed to create systems capable of generating high-quality, readable transcripts of human interactions. To achieve this, the dataset contains a broad array of accents, languages, ages, genders, and conversational topics.

The development of this dataset was initiated by the Linguistic Data Consortium (LDC), which created the Fisher collection protocol. Previous collection protocols, like CALLHOME, CALLFRIEND, and SwitchBoard-II, were more focused on language and speaker identification and lacked the diversity Fisher aimed to capture. The protocol concentrated on gathering numerous speakers to make short-duration calls with unfamiliar participants, maximizing inter-speaker variation and vocabulary. Although this approach slightly increased formality, it provided a vast diversity of demographic features, including age, gender, dialect region, English fluency, education level, occupation, and multilingual competence. Recruitment began in December 2002, using online advertisements and banners on computational linguistics and speech recognition platforms. The recruitment process spanned one year, during which participants' awareness of the collection's purpose and data processing policies was ensured. Participants were also financially compensated. The LDC initially set the topics to cover 40 themes, later expanding this set to 100 themes. Topics from the original set were reused extensively, while those added later were explored only once each. Ultimately, the LDC collected 16454 ten-minute calls, totaling 2742 hours of audio. Efforts were made to maintain gender balance, with females participating in 53% of the calls. A wide age distribution was achieved across the dataset, enriching it with a variety of regional pronunciations and non-US English varieties.

Two transcription methods were applied. The LDC utilized a variant of the Quick Transcription Specification (QTr), segmenting audio into 2–8 s chunks. Transcriptionists produced raw, verbatim transcripts without capitalization, punctuation, or markers for background noise and mispronunciations, though non-lexemes from a closed list were tagged, and acronyms were spelled out. The BBN/WordWave protocol transcribed entire conversations unmarred by segmentation, subsequently employing force alignment to timestamp each word. These methods yielded 200 hours of LDC-transcribed data and 1735 hours via BBN/WordWave.

The Fisher dataset has profoundly influenced ASR systems, facilitating their evolution towards handling spontaneous, diverse conversational data. Of course, the dataset exhibits some limitations. While striving for variety, its participant recruitment via self-selection means it might not represent broader speech patterns entirely. Formality increased somewhat due to structured topics, and while regional dialects are covered, there is an inherent limitation in an exhaustive representation of all speech idiosyncrasies. The dataset is accessible via the https://catalog.ldc.upenn.edu/LDC2004T19LDC website. Below we present a small excerpt of the Fisher dataset contents.

```
58.80 60.29 B: i saw that like a couple things
60.74 61.78 B: cause my roommate where watching it
62.06 65.03 A: yeah my roommates are you in college too
65.22 65.97 B: i'm in graduate school
66.05 68.06 A: oh yeah okay i just graduated from um
68.24 69.60 A: first time graduate last year
69.66 70.41 B: oh congrats
```

```
70.63 71.87 A: and how about what school are you in
71.82 72.40 B: i'm at fortum
72.78 73.94 A: oh wow so you're in new york
```

CALLHOME

The CALLHOME dataset [2], developed by the LDC, offers a valuable resource for the study of natural, unscripted speech. It comprises 120 telephone conversations, each lasting approximately 30 min, between native English speakers. 90 of these calls span international borders, originating from North America to various destinations outside the continent, while the remaining 30 occur within North America. Predominantly, these calls connect participants with family members or close friends, providing a genuine reflection of informal dialogue. The speakers involved in CALLHOME represent a spectrum of demographic backgrounds. Although participants were aware that their conversations were being recorded, they engaged freely on topics of their choosing, without being subject to specific constraints. This openness contributes to the authenticity of the dataset, capturing the nuances of everyday speech across different contexts.

The dataset totals 56.7 hours of audio divided into training, development, and test sets. Accompanying the audio are transcripts, detailing a continuous 5–10 min segment extracted from each recorded conversation. These transcripts account for 18.3 hours of transcribed audio and adhere to precise guidelines. Transcribers were instructed to focus on the middle portions of conversations and to faithfully preserve disfluencies, overlapping speech, and any mispronunciations, ensuring a precise representation of spontaneous speech. The CALLHOME corpus also includes anonymized speaker data. This information includes call IDs, gender, age, educational background, the state where the speaker was raised, and the initial digits of the dialed number's area code. While offering this demographic insight, the dataset maintains privacy by hiding any details regarding the interpersonal relationships among the speakers involved. Overall, the CALLHOME dataset is a rich repository of natural English language conversations, offering researchers extensive material for linguistic and computational analyses, aiding in the development of advanced speech recognition technologies.

SwitchBoard

The Switchboard Telephone Speech Corpus (LDC97S62) [5] is a historical dataset in the field of linguistic research. Originally collected by Texas Instruments between 1990 and 1991, this dataset was supported by DARPA and later released by NIST, with distribution managed by the Linguistic Data Consortium (LDC) in 1992–1993. Since its initial publication, significant corrections have been implemented, ensuring data integrity and consistency across subsequent iterations. Switchboard features around 2400 two-sided telephone conversations conducted by 543 speakers, consisting of 302 males and 241 females, representing diverse regions of the United States. A computer-controlled system facilitated these conversations, managing call initiation, topic introduction, and recording, ensuring separation into distinct audio channels for each speaker. The dataset includes discussions on approximately 70 topics, with 50 being particularly prominent. It was designed to ensure no pair of speakers conversed more than once and that topics were unique for each participant.

The corpus release, managed by LDC, involved rectifying errors from the original speech files. The revisions addressed speaker attribution inaccuracies and resolved missing files from the 1997 update. Further corrections were performed by The Institute for Signal and Information Processing (ISIP), resulting in enhanced transcriptions and word alignments, which are accessible for research purposes. Researchers have used the Switchboard corpus for multiple annotation projects, including discourse analysis, speech act tagging, part-of-speech tagging, and phonetic transcription. These projects have enriched the transcripts, incorporating comprehensive speaker attributes. In addition to the traditional transcripts, the Switchboard Dialog Act Corpus extends the dataset's usability through annotations of 60 basic dialogue act tags. These annotations draw upon the Discourse Annotation and Markup System of Labeling (DAMSL) framework, referred to as the SWBD-DAMSL labels. This labeling scheme covers socio-linguistic and theoretical discourse relations as well as formal models. The annotations describe 1155 conversations, each lasting approximately 5 min, including over 200000 utterances and totaling 1.4 million words.

Wall Street Journal

Developed to meet the demand for a large and diverse English language dataset, the WSJ corpus has been instrumental in training and evaluating early speech recognition models. At its inception, the WSJ corpus was notable for its size, comprising 400 hours of speech data alongside a large text dataset of 47 million words. The majority of the speech data consists of read articles from the Wall Street Journal, supplemented by a smaller section of spontaneous dictation. Designed with scalability in mind, the corpus accommodates various vocabulary sizes, ranging from 5000 to 64000 words or more. It supports both speaker-dependent and speaker-independent training paradigms. The WSJ corpus includes verbalized and non-verbalized punctuation, tailored speaker adaptation materials, and recordings conducted in moderately noisy environments.

The predominance of "read" speech in the WSJ corpus was a strategic choice to ensure cost-effectiveness. Speakers read from pre-prepared texts based on WSJ articles. To complement this, a limited amount of "spontaneous" speech was gathered. The text data within the WSJ corpus underwent preprocessing to prepare it for language model training. This involved converting numbers to orthographic forms, expanding abbreviations, and normalizing punctuation. The corpus features both "prompt" texts (what speakers read) and "truth" texts (used for training and evaluation), available in forms with both verbalized and non-verbalized punctuation. Despite its notable size at the time of release, the WSJ corpus faced challenges in achieving performance in speech recognition. The main drawback of the corpus is its reliance on read speech which does not encapsulate the subtleties of spontaneous conversations. The WSJ corpus marked a significant milestone in speech recognition research by providing a large, standardized dataset for the academic community. Although larger and more diverse datasets have since emerged, the WSJ corpus played an important role in the development and assessment of early speech recognition systems.

GigaSpeech

The GigaSpeech dataset [3] represents a major advancement in speech recognition. This large corpus addresses the limitations of older datasets such as Wall Street Journal, Switchboard, Fisher, and LibriSpeech, which have grown increasingly inadequate for advancing speech recognition technology due to their limited size, outdated content, and plateaued performance levels. With its vast scale and multi-domain approach, GigaSpeech sets a new standard for speech recognition datasets. GigaSpeech features an impressive 10000 hours of transcribed audio designed for supervised learning, along with an additional 40000 hours suitable for semi-supervised and unsupervised learning. The dataset contains a wide variety of speaking styles, including both read and spontaneous speech. The topics covered are diverse, ranging from arts and science. An interesting feature of GigaSpeech is its provision of both original and normalized transcription pairs, which facilitates the training of end-to-end systems with advanced post-processing abilities like punctuation and date or time normalization.

The collection process for GigaSpeech involved identifying 24 relevant categories from which content was sourced. This includes selecting audiobooks, podcasts with transcripts, and YouTube videos featuring closed captions within these categories. The annotation process was conducted through a multi-stage pipeline. It began with text normalization, followed by forced alignment using a Kaldi-based aligner. Audio was segmented according to silence and punctuation cues, and segment validation was performed to eliminate errors. Reference rewriting was used to correct common transcription mistakes, with a forced alignment graph applied to identify transcription errors and enable precise reference corrections for improved accuracy. The dataset is biased towards popular topics often found in YouTube and podcast selections. Additionally, there is a lack of control over recording conditions, which results in variability of audio quality. The dataset also does not include speaker demographics, which may limit research focused on speaker-specific traits.

GigaSpeech surpasses older corpora in several aspects. It offers a vastly greater scale, with over 10000 hours compared much smaller collections provided by older datasets. GigaSpeech is multi-source, drawing from audiobooks, podcasts, and YouTube, in contrast to the single-source nature of many earlier corpora. It contains diverse speaking styles, mixing read and spontaneous speech, and covers a broader range of topics beyond the traditional news, conversations, or audiobooks.

References

1. R. Ardila, M. Branson, K. Davis, M. Henretty, M. Kohler, J. Meyer, R. Morais, L. Saunders, F.M. Tyers, G. Weber, Common voice: A massively-multilingual speech corpus. arXiv preprint arXiv:1912.06670 (2019)
2. A. Canavan, D. Graff, G. Zipperlen, Callhome american english speech ldc97s42. Web Download (1997)
3. G. Chen, S. Chai, G. Wang, J. Du, W.-Q. Zhang, C. Weng, D. Su, D. Povey, J. Trmal, J. Zhang, et al., Gigaspeech: An evolving, multi-domain asr corpus with 10,000 hours of transcribed audio. arXiv preprint arXiv:2106.06909 (2021)
4. C. Cieri, D. Miller, K. Walker, The Fisher corpus: a resource for the next generations of speech-to-text. Int. Confer. Lang. Resour. Eval. **4**, 69–71 (2004)
5. J.J. Godfrey, E. Holliman, Switchboard-1 release 2 ldc97s62. Web Download (1993)
6. F. Hernandez, V. Nguyen, S. Ghannay, N. Tomashenko, Y. Esteve, Ted-lium 3: Twice as much data and corpus repartition for experiments on speaker adaptation, in *Speech and Computer: 20th International Conference, SPECOM 2018, Leipzig, Germany, September 18–22, 2018, Proceedings 20* (Springer, 2018), pp. 198–208
7. A. Nagrani, J.S. Chung, W. Xie, A. Zisserman, Voxceleb: Large-scale speaker verification in the wild. Comput. Speech Lang. **60**, 101027 (2020)
8. V. Panayotov, G. Chen, D. Povey, S. Khudanpur, Librispeech: an asr corpus based on public domain audio books, in *2015 IEEE International Conference on Acoustics, Speech and Signal Processing (ICASSP)* (IEEE, 2015), pp. 5206–5210
9. D. Povey, A. Ghoshal, G. Boulianne, L. Burget, O. Glembek, N. Goel, M. Hannemann, P. Motlicek, Y. Qian, P. Schwarz, et al., The Kaldi speech recognition toolkit, in *IEEE Workshop on Automatic Speech Recognition and Understanding*, number EPFL-CONF-192584 (IEEE Signal Processing Society, 2011)
10. T.F. Smith, M.S. Waterman et al., Identification of common molecular subsequences. J. Mol. Biol. **147**(1), 195–197 (1981)
11. I.H. Witten, T.C. Bell, The zero-frequency problem: Estimating the probabilities of novel events in adaptive text compression. IEEE Trans. Inf. Theory **37**(4), 1085–1094 (1991)
12. P. Szymański, L. Augustyniak, M. Morzy, A. Szymczak, K. Surdyk, P. Żelasko, Why aren't we ner yet? artifacts of asr errors in named entity recognition in spontaneous speech transcripts, in *Proceedings of the 61st Annual Meeting of the Association for Computational Linguistics (Volume 1: Long Papers)* (2023), pp. 1746–1761

GPSR Compliance
The European Union's (EU) General Product Safety Regulation (GPSR) is a set of rules that requires consumer products to be safe and our obligations to ensure this.

If you have any concerns about our products, you can contact us on

ProductSafety@springernature.com

In case Publisher is established outside the EU, the EU authorized representative is:

Springer Nature Customer Service Center GmbH
Europaplatz 3
69115 Heidelberg, Germany

www.ingramcontent.com/pod-product-compliance
Lightning Source LLC
Chambersburg PA
CBHW061253161025
33954CB00068B/77
* 9 7 8 3 0 3 1 8 8 5 6 5 5 *